Michael Schulz (Ed.)
Ion-Atom Collisions

Also of Interest

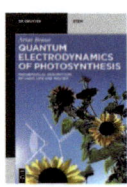

Quantum Electrodynamics of Photosynthesis
Mathematical Description of Light, Life and Matter
Braun, 2020
ISBN 978-3-11-062692-6, e-ISBN 978-3-11-062994-1

X-ray Studies on Electrochemical Systems
Synchrotron Methods for Energy Materials
Braun, 2017
ISBN 978-3-11-043750-8, e-ISBN 978-3-11-042788-2

Radiative Neutron Capture
Primordial Nucleosynthesis of the Universe
Dubovichenko, 2019
ISBN 978-3-11-061784-9, e-ISBN 978-3-11-061960-7

On the Origin of Natural Constants
Axiomatic Ideas with References to the Measurable Reality
Good, 2018
ISBN 978-3-11-061028-4, e-ISBN 978-3-11-061238-7

Ion-Atom Collisions

The Few-Body Problem in Dynamic Systems

Edited by
Michael Schulz

DE GRUYTER

Editor
Dr. Michael Schulz
Department of Physics
Missouri S. T.
1315 N. Pine St. 65409 Rolla, MO
USA
schulz@mst.edu

ISBN 978-3-11-057942-0
e-ISBN (PDF) 978-3-11-058029-7
e-ISBN (EPUB) 978-3-11-057946-8

Library of Congress Control Number: 2019947517

Bibliographic information published by the Deutsche Nationalbibliothek
The Deutsche Nationalbibliothek lists this publication in the Deutsche Nationalbibliografie;
detailed bibliographic data are available on the Internet at http://dnb.dnb.de.

© 2019 Walter de Gruyter GmbH, Berlin/Boston
Cover image: EzumeImages / iStock / Getty Images Plus
Typesetting: VTeX UAB, Lithuania
Printing and binding: CPI books GmbH, Leck

www.degruyter.com

Preface

Research on ion-atom collisions plays a crucial role in advancing our understanding of the fundamentally important few-body problem. It is therefore not surprising that a rich literature in scientific journals on these research activities, usually addressing readers with expertise in the field, has emerged over the last several decades. However, it is surprising that more comprehensive publications, addressing a more general readership, are rather sparse. Although numerous review articles on various topics in ion-atom collisions are available, these serve mostly as a reference for researchers in the field. Furthermore, several excellent textbooks on ion-atom scattering theory have been published. These books take a rather formal approach and assume that the reader has already been thoroughly introduced into the basic concepts and ideas of ion-atom collisions. Books offering a general introduction, covering both experimental and theoretical approaches, are to a large extent missing. The present book aims to provide such an introduction. As such, it is intended for a readership within a broad range of background knowledge in the field, ranging from senior researchers to postdocs, graduate students, and even advanced undergraduate students with some research experience. It makes no claim to be complete; in view of the size of the field this would be an impossible task. However, we hope that it provides a sufficiently comprehensive overview of the research activities to give the reader a solid idea about the most important goals of and methods in ion-atom collision research.

https://doi.org/10.1515/9783110580297-201

Contents

List of Contributing Authors

R. O. Barrachina
Centro Atómico Bariloche and Instituto Balseiro
Comisión Nacional de Energía Atómica (CNEA),
Universidad Nacional de Cuyo and Consejo Na-
cional de Investigaciones Científicas y Técnicas
(CONICET)
Av. Bustillo 9500
8400 San Carlos de Bariloche
Argentina

A. Cassimi
CIMAP, CEA-CNRS-ENSICAEN-UNICAEN
Normandie Université
BP5133
F-14050 Caen Cedex 04
France

M. F. Ciappina
Institute of Physics of the ASCR
ELI-Beamlines
Na Slovance 2
182 21 Prague
Czech Republic

Alain Dubois
Sorbonne Université, CNRS
Laboratoire de Chimie Physique-Matière et Ray-
onnement
F-75005 Paris
France

Daniel Fischer
Missouri University of Science & Technology
Physics Dept.
1315 N. Pine St.
Rolla, MO 65409
USA

X. Fléchard
ENSICAEN, UNICAEN, CNRS/IN2P3, LPC Caen
Normandie Univ
14000 Caen
France

B. Gervais
CIMAP, CEA-CNRS-ENSICAEN-UNICAEN
Normandie Université
BP5133
F-14050 Caen Cedex 04
France

Tom Kirchner
Department of Physics and Astronomy
York University
Toronto, Ontario M3J 1P3
Canada

A. Méry
CIMAP, CEA-CNRS-ENSICAEN-UNICAEN
Normandie Université
BP5133
F-14050 Caen Cedex 04
France

F. Navarrete
Department of Physics
Kansas State University
Manhattan, KS 66506
USA

J. Rangama
CIMAP, CEA-CNRS-ENSICAEN-UNICAEN
Normandie Université
BP5133
F-14050 Caen Cedex 04
France

Reinhold Schuch
Physics Department
Stockholm University, AlbaNova
10691 Stockholm
Sweden

Michael Schulz
Missouri University of Science & Technology
Physics Dept.
1315 N. Pine St.
Rolla, MO 65409
USA

Nicolas Sisourat
Sorbonne Université, CNRS
Laboratoire de Chimie Physique-Matière et Ray-
onnement
F-75005 Paris
France

A. B. Voitkiv
Institut für Theoretische Physik I
Heinrich Heine University of Düsseldorf
Universitätsstrasse 1
40225 Düsseldorf
Germany

Michael Schulz

1 General considerations

1 Few-body problem

The famous Rutherford experiment [1], which can be considered the birth of ion-atom collision research, is of fundamental importance for the development of modern physics for several reasons. It paved the way for a solid understanding of atomic structure. Initially, a "planetary" atomic model emerged from the experimental observations which, however, could not explain the stability of atoms (classically, the electrons should spiral and crash into the nucleus). But with the development of quantum-mechanics, this problem was resolved and atomic structure can now be regarded as being conceptually understood. Furthermore, the Rutherford experiment marked the splitting of the physics of the atom into two branches, which we now call nuclear and atomic physics. Nuclear physics later led to a new field, to which we now refer as elementary particle or high energy physics. The research developments yielding the exciting insights into the subatomic world of quarks, quantum-field effects, the fundamental forces, etc. [2], which were achieved over the years, are not conceivable without the Rutherford experiment.

Although the atomic physics branch emerging from Rutherford's studies resulted in a solid understanding of atomic structure within a relatively short time [3], research in atomic physics nevertheless continues to address fundamental questions in science. More specifically, atomic collision physics is of high relevance to one of the most fundamentally important and yet unsolved problems in physics, the few-body problem (FBP) [4, 5]. To illustrate the FBP and its significance, we address the very basic question "what does it take to understand Nature?" Crudely speaking, the answer to this question can be summarized in two points. First, we need to obtain a solid concept on the forces acting in Nature on an elementary particle level. This point falls within the domain of quantum field theory. Second, we need to solve the FBP, which is a multidisciplinary task. However, we will argue that atomic collision physics takes a particularly important role in this task.

In our current understanding, we need at most four fundamental forces to describe Nature, which are gravity, the weak force, the electromagnetic force, and the strong force.[1] All of these forces are mediated by the interacting particles exchanging another particle characteristic to the specific force, the so-called gauge bosons.

---

[1] Accounting for the unification of the weak and electromagnetic forces at most three forces are required.

Michael Schulz, Missouri University of Science & Technology, Physics Dept., 1315 N. Pine St., Rolla, MO 65409, USA, e-mail: schulz@mst.edu

https://doi.org/10.1515/9783110580297-001

Since the gauge boson can only be emitted and absorbed by one particle at a time, the mediation of a force is fundamentally a two-body process, i. e., forces can only act within pairs of particles. This directly leads to the question of how do systems containing more than two particles evolve under the influence of these pairwise acting forces? This question poses an enormous challenge because the Schrödinger equation (or Dirac equation in relativistic cases) is not analytically solvable for more than two mutually interacting particles, even when the underlying force(s) are precisely known. This is the essence of the FBP. Because of this difficulty, theory has to resort to numeric approaches.

Properties of stationary atoms, like energy levels, can usually be calculated with high precision using numeric methods like the multiconfiguration Hartree–Fock procedure [3]. However, the FBP represents a major challenge to the theoretical description of dynamic systems like e. g. collisions. Detailed experimental data are needed on such systems to guide theoretical modelling efforts. For two reasons, atomic collisions are particularly suitable to study the dynamic FBP: first, the fundamental force underlying atomic systems, the electromagnetic force, is essentially understood. Therefore, experiments on atomic systems directly test the theoretical description of the few-body dynamics. In contrast, in the case of, e. g., nuclear systems it is not clear whether experiments test the theoretical description of the few-body dynamics or of the underlying fundamental force, i. e., the strong force, which is not nearly as well understood as the electromagnetic force.

Second, in atomic collisions systems with small particle numbers (of the order 3 to 5) can be studied. With the development of recoil-ion momentum spectroscopy [6, 7] (see Chapters 2, 6, and 9), it has become feasible to perform experiments for such small particle numbers in which the complete kinematic information about every single particle in the system is extracted (kinematically complete experiments) [8]. In contrast in, e. g., solid state physics the system under investigation typically contains a huge particle number (of the order of Avogadro's number), for which a kinematically complete experiment is not possible. In this case, only statistically averaged or collective quantities can be obtained, which do not offer a very sensitive test of theory. A potential lack of understanding of the few-body dynamics (for such large particle numbers it is actually more appropriate to talk about many-body dynamics) of the system could simply be hidden in the statistics.

2 Processes occurring in ion-atom collisions

The few-body dynamics in ion-atom collisions can vary significantly depending on the specific process occurring in the scattering event. In the Rutherford experiment, elastic scattering between α-particles and Au nuclei was studied [1]. In the analysis of the results of this experiment, Rutherford completely ignored the presence of a large number of electrons in the Au-atom and assumed that (except for very small scattering angles) the deflection of the projectile is entirely due to the Coulomb force between the two nuclei. In contrast, in modern ion-atom collision experiments the interest is focused on processes in which in addition to the deflection of the projectile also a transition of one or more electrons to a different state of the target or the projectile occurs. Such processes usually lead to a change of the internal energy of the collision partners and are therefore inelastic. Atomic scattering processes typically lead to scattering angles of a small fraction of a degree. In this angular range, the Coulomb force between the electron and the projectile can make a major contribution to the projectile deflection and is often even dominant over the Coulomb repulsion between the two nuclei.

The most basic inelastic processes in ion-atom collisions are those in which only a single transition of one electron occurs. These are excitation, ionization, and electron capture and for a projectile with initial charge Q+ colliding with a neutral target B can be represented by the following reaction equations:

$$A^{Q+} + B \rightarrow A^{Q+} + B^* \tag{1.1a}$$

$$A^{Q+} + B \rightarrow A^{Q+} + B^+ + e^- \tag{1.1b}$$

$$A^{Q+} + B \rightarrow A^{(Q-1)+} + B^+ \tag{1.1c}$$

Of course, if $Q < Z_p$, where Z_p is the atomic number of the projectile, electron transitions resulting in excitation or ionization can also occur in the projectile, where the latter process is then usually referred to as electron loss.

The few-body dynamics does not only sensitively depend on the specific process occurring in the collision, but also on parameters characterizing the collision system like, e. g., the projectile speed and the nuclear charges of the collision partners. Two particularly useful quantities are the projectile charge-to-speed ratio $\eta = Q_p/v_p$, called the perturbation or Sommerfeld parameter, and the ratio between the projectile and target nuclear charge $\lambda = Z_p/Z_t$. As a crude rule of thumb for small η, the collision dynamics tends to be dominated by a single interaction of the projectile nucleus with the active target electron and with increasing η multiple interactions become increasingly important. Therefore, for small η low-order perturbation theory is often an adequate approach to describe the process of interest. With increasing η, convergence of the perturbation expansion becomes slower, i. e., an increasing number of higher-order expansion terms are required to achieve satisfactory accuracy, and eventually the expansion may not converge at all. Therefore, for η approaching unity (in atomic units) nonperturbative approaches are more promising.

For η significantly larger than 1 and $\lambda \approx 1$, treating the projectile potential as a perturbation to the target system is no longer useful even for a qualitative description of the few-body dynamics. Under these conditions, the collision time is long enough for the electrons, especially inner-shell electrons, to adjust to the two-center potential generated by the nuclei of both collision partners. An inelastic collision process is then best viewed as an electronic transition between quasi-molecular states formed by the two-center potential. These quasi-molecular states asymptotically turn into atomic states of the separated atoms at very large distances between the collision partners and of the united atom with a nuclear charge $Z_u = Z_p + Z_t$ at very small distances. By studying electron transitions between such quasi-molecular states it is possible to perform spectroscopy on the united atom [9]. One can even envision doing spectroscopy of super-heavy ions with a nuclear charge of around 150 and higher, i. e., on ions which do not even exist as stationary particles.

In this regime of very large η, the sign of the projectile charge makes a large qualitative difference on the few-body dynamics. Therefore, collisions with anti-protons have attracted considerable interest [10] (also see Chapter 3). Because of the negative projectile charge, the force between the projectile and the active electron is repulsive. As a result, the two-center potential formed by the anti-proton and the target nucleus does not support the formation of quasi-molecular states. The same two-center potential is also formed in electron collisions with atoms. However, here the values of η that can be realized are limited by the threshold energy for the process of interest. For example, ionization of atomic hydrogen requires a minimum center-of-mass energy in the collision of 13.6 eV (the ionization potential of H). This energy corresponds to a projectile speed of 1 a. u. so that the maximum η for ionization of H by electron impact is 1. For anti-protons, the minimum speed is smaller by about three orders of magnitude due to the much larger mass so that in this case there is practically no limitation on η. Therefore, anti-protons are very well suited to study emission of electrons from a single center at very large perturbations (see Chapter 3).

In one-electron processes, the presence of any passive electrons often does not have a large effect on the dynamics of the transition of the active electron, but rather it is sufficiently well described in terms of a screening function which leads to a reduced, distance-dependent effective nuclear charge. In fact, just incorporating the passive electrons in terms of a constant effective nuclear charge qualitatively often yields satisfactory results. However, in processes where two or more electrons undergo a transition, the interaction between both active electrons can play an active and important role [11]. For example, double ionization of the target can proceed by the projectile ejecting each electron by independently interacting with both of them but especially at large projectile energies, another mechanism tends to be more important. Here, the projectile only interacts with one electron while the second electron is ejected to the continuum by an interaction with the first ejected electron. Here, the transitions of both electrons are correlated with each other (electron-electron correlation). Similar channels can also be important in other two- (or many-) electron processes, like double capture, double excitation, transfer-ionization, etc. In studies of processes involving such multiple electronic transitions, the interest is often focused on electron-electron correlation effects, which results in qualitatively different few-body dynamics than in one-electron processes. However, in this book we focus on single-electron transitions.

Finally, for molecular targets the presence of multiple nuclei opens reaction channels which do not exist for atomic targets. For example, the collision can lead to an electron transition to a repulsive state, i. e., the molecule will dissociate. Furthermore, the nuclei of the molecule represent multiple scattering centers to the projectile. Therefore, indistinguishable diffraction of the incoming projectile wave from these scattering centers can lead to interference patterns in measured spectra. The phase angle in the interference term tends to be quite sensitive to the details of the few-body collision dynamics. As a result, the analysis of such interference patterns is particularly promising in attempts to advance our understanding of the FBP.

All of the processes mentioned here have been studied extensively, both experimentally and theoretically, and a broad selection will be reviewed and analyzed in more detail in this book. In most collision studies the cross-section for the process of interest is measured or calculated. Therefore, in the following a basic introduction to the concept of a scattering cross-section is given.

3 The scattering cross-section

The objective of a scattering experiment is often to determine the probability with which a certain process or a specific final state of the collision occurs. However, this probability depends on the experimental set up, more specifically, on the thickness and density of the target. The scattering cross-section σ is a useful quantity which is proportional to the reaction probability, but at the same time independent of the experimental preparation of the target.

To understand the connection between σ, which is an area, and the reaction probability, consider a target filling a cuboid of thickness l and a front surface area A. For simplicity, let us first consider a target consisting of N billiard balls of radius R, so that their cross-sectional area $\sigma = \pi R^2$, spread randomly over the entire cuboid, as shown in Figure 1.1. We make the assumptions that $N\sigma \ll A$, so that the probability that one billiard ball is (partly) covered by others is negligible. Suppose that marbles, with a size small compared to the billiard balls, are thrown with a speed v_0 perpendicular to the front surface, but otherwise randomly (i. e., without "aiming" at any billiard ball), into the cuboid. We can ask for the probability P that the marbles get deflected by one of the billiard balls (with the assumption $N\sigma \ll A$ the probability for multiple deflections is negligible). In this example, deflection requires physical contact between the marble and one of the billiard balls. Therefore, P is given by the ratio of the area covered by billiard balls to the total front surface area A of the cuboid:

$$P = N\sigma/A = nAl\sigma/A = nl\sigma \qquad (1.2)$$

where n is the number density of billiard balls $n = N/V$. Although, the product nl is the areal target density, it is often referred to as the target thickness. Equation (1.2) shows that the cross-section is a quantity that is directly proportional to the reaction probability and which is independent of the target thickness.

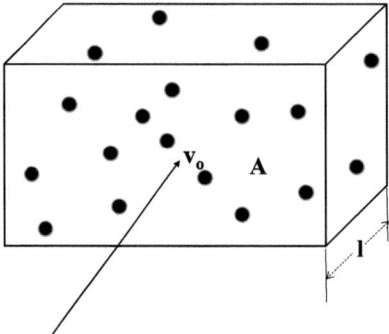

Figure 1.1: Illustration of the scattering cross-section for classical marble—billiard ball collisions. Billiard balls are randomly distributed in a cuboid of front surface area A and thickness l. Marbles pass through the cuboid with speed v_0. The probability for deflection of marbles by one of the billiard balls is given by the ratio of the area covered by billiards ball to the total front surface area.

Returning to collisions between ions and atoms two important differences to classical collisions between marbles and billiard balls should be noticed: first, because the Coulomb force is a remote long-range force the deflection of the projectiles does not require "physical contact" with the target. Second, the projectile approaching a target atom within a certain maximum distance is no guarantee that any specific inelastic process will occur, while in our classical example physical contact will guarantee deflection of the marble. With these differences in mind, we can define the scattering cross-section as follows: the scattering cross-section is the *effective* area within which the projectile needs to approach the target atom for the process of interest (e. g., deflection or ionization of the target) to occur.

In an experiment studying the total cross-section, the event rate at which the process of interest occurs has to be determined. For example, to measure the total cross-section for ionization of the target one can count events in which the charge state of the target is increased by one unit and at the same time the projectile charge remains unchanged. However, it is possible to extract more detailed information from the experiment. For example, instead of counting all events characterizing ionization in terms of the projectile and target charge, one could count only those events in which the projectiles got deflected at a specific scattering angle. This rate is a measure of the differential cross-section $d\sigma(\theta_p, \varphi_p)/d\Omega_p$, where the solid angle $d\Omega_p$ subtended by the projectile detector is related to the scattering angle by

$$dΩ_p = \sin θ_p dθ_p dφ_p \qquad (1.3)$$

Here, $_p$ and $θ_p$ are the polar and azimuthal angular components of the final projectile momentum.

The level of detail can be further enhanced by experimentally determining additional kinematic quantities of the particles involved in the final state of the collision like, e. g., the ejection angle of any emitted electron. The kinematics of the collision is completely characterized by the momentum vectors of the unbound particles emerging from the collision, which are often expressed in spherical coordinates $(E, θ, φ)$, where the kinetic energy E is equivalent to the magnitude of the momentum p. Therefore, if the final state of the collision contains n unbound particles then $3n$ momentum components have to be determined. However, not all of these components are independent of each other because of total momentum and energy conservation. As a result, the direct measurement of $3n - 4$ momentum components already represents a complete characterization of the collision kinematics and such an experiment is called kinematically complete. The direct measurement of any additional momentum component does not provide any additional information about the collision process. Therefore, the maximum degree of differentiality of a cross-section for a process involving n particles is given by $3n - 4$ and this is called a fully differential cross-section (FDCS). Here, we are neglecting spin-changing transitions, which tend to be very weak in ion-atom collisions.

An FDCS can be represented as a function of any set of $3n - 4$ momentum components fully determining the kinematics. For example, in the case of single target ionization, which involves three unbound particles in the final state (the scattered projectile, the ejected electron, and the recoiling target ion), it is very common to present cross-sections differential in the solid angle $dΩ$ of two particles and in the energy dE of one particle. At first glance, this may seem to represent a triple differential cross-section and indeed this notation is widespread in the literature on single ionization. However, the right-hand side of equation (1.3) shows that the solid angle is really a double differential (in $θ$ and $φ$). Therefore, a FDCS for $n = 3$ is fivefold differential, in accordance with $3n - 4$.

In the case of $n = 2$ like, e. g., in the Rutherford experiment on elastic scattering, a kinematically complete experiment requires the direct measurement of two momentum components. Here, the cross-section $dσ(θ)/dΩ$ is already fully differential. For single target excitation and electron capture (i. e., processes (1.1a) and (1.1c)) $n = 2$ as well. However, here the cross-section does not only depend on the kinematics, but also on the initial and final state of the electronic transition. For these processes, the most detailed data that can be extracted from an experiment are state-selective differential cross-sections $dσ(θ)/dΩ$. The initial and/or final state of the electronic transition can be determined, e. g., from the energy loss of the projectile [12], which is a kinematic parameter. However, the determination of the initial and/or finals states of the transition is not considered as increasing the degree of differentiality because both states have discrete energy levels.

In this book, we will review studies on some processes mentioned above. Different experimental and theoretical techniques will be covered. Depending on the process under investigation and on the goals of the study it may be more desirable or feasible to analyze total or differential cross-sections. The most sensitive test of theoretical models by experiment is offered by FDCS because the integration over kinematic parameters, resulting in less differential or total cross-sections, can (partly) "average out" discrepancies between measured and calculated cross-sections. However, the disadvantage of FDCS is that data are only obtained for a very restricted range of kinematic settings, i. e., FDCS are not comprehensive. A total cross-section or single differential measurement can provide information as to which process is dominant in a specific collision system and which mechanism is dominant in leading to a specific process, while fully differential measurements can provide this information only for very specific conditions. Furthermore, experimental studies of FDCS (or even less differential cross-sections) are not always feasible, e. g., when the cross-section for the process of interest is too small, the process cannot be identified with sufficient detection probability, or the combination of projectile beam intensity and target density cannot be made sufficiently large. Therefore, (multiple) differential and total cross-section studies complement each other to provide a comprehensive picture of the few-body dynamics in ion-atom collisions.

The experimental methods presented in this book range from single particle detection in merged beam experiments, where differential measurements are extremely challenging, to multiparticle detection yielding multiple differential cross-sections including FDCS, to ion storage rings, which due to the large beam intensity and high beam quality make experiments possible which are not feasible with "conventional" accelerators. Theoretical models cover both perturbative and nonperturbative approaches.

Bibliography

[1]   Rutherford E. Philos. Mag. 1911;21:669.
[2]   Halzen F, Martin AD. Quarks and Leptons. New York: John Wiley and Sons; 1984.
[3]   Froese-Fischer C, Brage T, Johnsson P. Computational Atomic Structure: An MCHF Approach, CRC Press (1997).
[4]   Schulz M, Moshammer R, Fischer D, Kollmus H, Madison DH, Jones S, Ullrich J. Nature 2003;422:48.
[5]   Rescigno T, Baertschy M, Isaacs WA, McCurdy CE. Science 1999;286:2474.
[6]   Ullrich J, Moshammer R, Dorn A, Dörner R, Schmidt L, Schmidt-Böcking H. Rep. Prog. Phys. 2003;66:1463.
[7]   Dörner R, Mergel V, Jagutzski O, Spielberger L, Ullrich J, Moshammer R, Schmidt-Böcking H. Phys. Rep. 2000;330:95.

[8]   Schulz M, Madison DH. International Journal of Modern Physics A. 2006;21:3649.
[9]   Schuch R, Ingwersen H, Justiniano E, Schmidt-Böcking H, Schulz M, Ziegler F. J. Phys. B. 1984;17:2319.
[10] Kirchner T, Knudsen H. J. Phys. B. 2011;44:122001.
[11] McGuire JH. Electron Correlation Dynamics in Atomic Collisions. Cambridge University Press; 2016.
[12] Park JT. Adv. At. Mol. Phys. 1983;19:67.

Michael Schulz

2 Ionization: three-body scattering

1 Introduction

With regard to the few-body problem ionization takes a special role among the basic inelastic atomic collision processes because the final state involves three unbound particles, the scattered projectile, the residual target ion (or recoil ion) and the ejected electron. In contrast, kinematically, both excitation and electron capture represent two-body scattering processes. Experimental studies on ionization therefore offer a particularly sensitive test of the theoretical description of the few-body dynamics. As a result, a rich literature on both experimental and theoretical studies on ionization exist, which range from total cross sections (for reviews see, e. g., [1, 2]) to single and double differential cross sections (for reviews see, e. g., [2, 3]) to fully differential cross-sections (for a review see, e. g., [4]). In the following, some of the most important results will be summarized and discussed.

2 Total cross-sections

Compared to (multiple) differential cross-sections total cross-sections have the disadvantage that they lack the detail offered by the former so that they are not ideally suited to sensitively test theoretical calculations. However, the big advantage of total cross-section studies is that they provide accurate information about the relative importance of various reaction mechanisms, while differential cross-sections provide this information only for specific kinematic conditions. Therefore, total and differential studies complement each other and both are essential for a thorough understanding of the few-body reaction dynamics.

The reaction equation for ionization (equation (1.1b) in Chapter 1) shows that there are various possibilities of identifying ionization. One method is to measure the intensity of positively charged target ions produced in the collision. However, such a target ion will also be produced by a capture process (equation (1.1c) in Chapter 1). Therefore, the intensity of the ejected electrons has to be measured simultaneously with the ion intensity. This approach was successfully applied for the first time by Keene [5] and further developed by several groups [e. g., 6–9]. In principle, the detection of the ejected electrons alone should be sufficient to identify ionization. One major difficulty is that the measurement of the electron current is usually afflicted with a large background of electrons not originating from the collision under investigation.

Michael Schulz, Missouri University of Science & Technology, Physics Dept., 1315 N. Pine St., Rolla, MO 65409, USA, e-mail: schulz@mst.edu

https://doi.org/10.1515/9783110580297-002

However, with high resolution energy-analyzers electrons can be detected with a sufficiently low level of background, which can be determined in a separate measurement (without target) and subtracted from the data. If these measurements are performed over a sufficiently large range of electron energies and ejection angles, the total cross-section can be obtained by integration over both parameters [10, 11]. Finally, the flux of projectiles that cause target ionization can be determined. However, it is not sufficient to simply count projectiles which did not change the charge state in the collision because this flux would be dominated by projectiles which merely got elastically scattered from the target without causing ionization. To select ionization events, the energy loss of the projectiles has to be measured as ionization has a threshold energy which is equal to the ionization potential of the target. This method was pursued by Park et al. [12].

In Fig. 2.1 total cross sections for ionization of atomic hydrogen by anti-proton impact are shown as a function of the projectile energy [13, 14]. In the present context, we do not need to be concerned with the differences between proton and anti-proton impact. Such differences are very important at small projectile energies and are discussed in detail in Chapter 3. Here, we focus on larger projectile energies where the reaction dynamics for proton and antiproton impact are qualitatively similar.

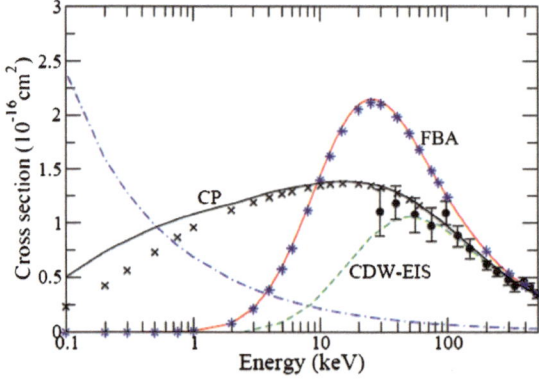

Figure 2.1: Total cross-sections for ionization of atomic hydrogen by antiproton impact as a function of projectile energy. The red curve shows a perturbative first-order calculation (FBA), the green curve a perturbative higher-order calculation (CDW-EIS), and the black curve a nonperturbative higher-order calculation (CP).

The comparison between experiment and theory provides some important insight into the collision dynamics. The red, green, and black curves represent calculations based on the first Born approximation (FBA), on the continuum distorted wave — eikonal initial state model (CDW-EIS), and the coupled pseudo-states approach (CP) [13]. The crosses and the blue curve are not needed in the present context. Neither is it important for the following analysis to fully understand the numeric details of these calculations; these will be discussed in Chapters 3, 5, and 7. For now, it suffices to state that the FBA and CDW-EIS approaches represent perturbative first- and higher-order calculations, respectively, and the CP model a nonperturbative calculation which conceptually corresponds to an infinite-order calculation. Here, first-order means that theory only accounts for a single interaction between the projectile and the active electron. The interaction between the two nuclei of the collision system (NN interaction) as well as multiple interactions between the projectile and the electron are completely omitted. In contrast, the CDW-EIS model accounts for the NN interaction, following or preceding the primary projectile-electron interaction, as well as for multiple projectile-electron interactions in the initial and final state wavefunction of the collision system.

One important observation is that while all three calculations differ from each other at small energy they approach each other with increasing energy and become essentially identical at about 400 keV. Furthermore, at this energy the calculations are in nearly perfect agreement with the experimental data, while at smaller energies the measurements are only reproduced by the two higher-order calculations. This observation shows that at large energies higher-order contributions are negligible compared to the first-order mechanism, i. e., even the NN interaction is completely insignificant. At first glance, this may seem like a surprising result because the Rutherford cross-section for elastic scattering is exclusively based on the NN interaction. However, two important differences between Rutherford scattering and ionization should be emphasized: first, in the original Rutherford experiment relatively large scattering angles (a few degrees up to backscattering) were investigated. Classically, such large angles cannot be realized by the projectile—electron interaction due to the very large projectile to electron mass ratio. Second, ionization cannot proceed without the electron interacting with the projectile because the latter needs to provide the energy to lift the electron to the continuum. Therefore, the NN interaction can only participate within higher-order processes, which, in turn, generally become less important with increasing energy because of the decreasing collision time. This is a common feature of inelastic atomic collisions: the collision dynamics underlying total cross-sections tend to be dominated by the projectile-electron interaction, especially at large energies. It should be noted that apart from the energy the relative importance of such higher-order contributions also depends on the projectile charge state. A useful parameter to characterize the collision with regard to higher-order contributions is the projectile charge to speed ratio η known as the perturbation or Sommerfeld parameter. First-order perturbation theory is usually deemed justified for $\eta \ll 1$.

3 Differential ejected electron spectra

The energy dependence of the total cross-sections shows that for antiproton—H collisions higher-order contributions to ionization become significant below approximately 200 keV (corresponding to $\eta > 0.35$). However, this energy dependence does not exhibit much structure and it is therefore difficult to determine what types of higher-order mechanisms are particularly important. Further information about the few-body dynamics is obtained by recording the energy spectra of the ejected electrons. Usually, electron spectrometers are equipped with narrow entrance and exit slits in order to achieve sufficient energy resolution. This means that only electrons ejected in a certain direction are energy-analyzed so that the electron flux on the detector reflects a double differential (in energy and solid angle of the electrons) ionization cross-section (DDCS).[1]

In Figure 2.2 DDCS for 30 MeV C^{6+} + H_2 collisions are shown for electron emission angles ranging from 75° to 105° as a function of electron energy [15]. Just like the projectile energy dependence of the total cross-sections the DDCS, too, do not reveal any structures that could be interpreted as signatures of significant higher-order contributions. Nevertheless, comparison to theory clearly demonstrates that such contributions are very important for this collision system. The dotted curves show calculations based on the FBA, which only accounts for first-order contributions. The solid line represents a CDW-EIS calculation, which conceptually includes higher-order contributions in the final (and to a lesser extent initial) state wavefunction. The two calculations differ from each other by as much as nearly an order of magnitude and the CDW-EIS calculations reproduce the experimental data almost perfectly (except for large electron energies). This shows that indeed higher-order contributions are very important. Considering the large value of η for this collision system ($\eta = 0.6$) this finding is not surprising.

One important question is what type of higher-order ionization mechanisms mostly contribute to the DDCS. As a first step, we will address this question by crudely distinguishing between higher-order processes involving the NN interaction and those involving multiple interactions between the projectile and the active electron. The CDW-EIS calculation shown in Figure 2.2 represents a relatively early version of this model and it did not account for the NN interaction at all. Therefore, the differences between the FBA and CDW-EIS calculations are entirely due to multiple projectile-electron interactions. Effects due to the NN interaction are either insignificant or (almost) not observable in differential ejected electron spectra. The latter point will be discussed in more detail in the next section. Here, we only mention that later CDW-EIS models were developed which did include the NN interaction [e. g. 16–21, see also Chapter 5].

Theoretical DDCS have been tested even under extreme conditions of very large η. As an example, Figure 2.3 shows DDCS for ionization of helium by 3.6 MeV/amu Au^{53+}

---

1 In a strict sense, it is actually a triple differential cross-section because the solid angle is a double differential (see Chapter 1). However, in the literature it is common to treat the solid angle as a single differential.

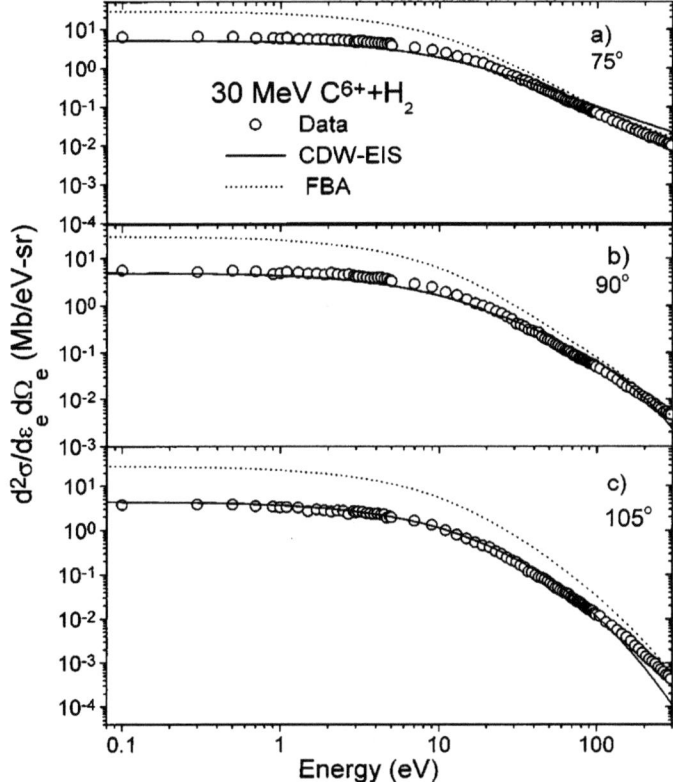

Figure 2.2: Double differential cross-sections, in ejected electron energy and electron solid angle, for ionization of H_2 by 30 MeV C^{6+} impact for various fixed ejection angles as a function of electron energy. The dotted and solid curves represent the FBA and CDW-EIS models, respectively.

impact [22]. The data are presented in a different format compared to those of Figure 2.2: instead of plotting the DDCS for fixed electron angles as a function of electron energy they are shown for fixed transverse electron momentum components as a function of the longitudinal (i. e., in the projectile beam direction) component. It should be noted that these presentations are unambiguously related to each other by Jacobi transforms, i. e., they are equivalent. The solid curves in Figure 2.3 show CDW-EIS calculations, which are conceptually equivalent to those plotted in Figure 2.2. This collision system corresponds to $\eta = 4.4$ and it does not appear to be very realistic to treat the projectile as a small perturbation to the target Hamiltonian. It is therefore quite remarkable that the data are well reproduced by theory. This good agreement reinforces that the NN interaction (not included in the calculations) has no significant observable effect on the double differential electron spectra, even for extremely large η. It also illustrates one important advantage of CDW-EIS over other higher-order perturbative approaches like, e. g., the Born series. In the latter, the transition amplitude

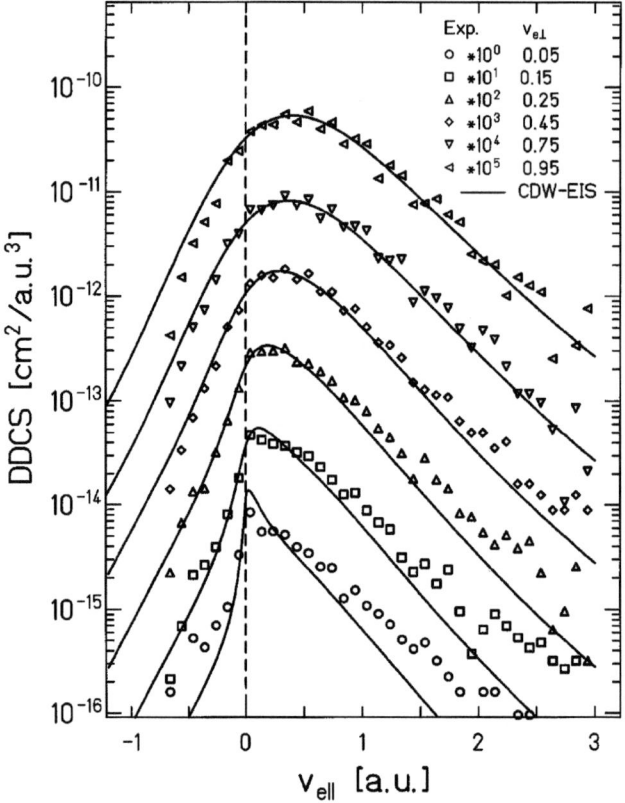

Figure 2.3: Double differential cross-sections, in the transverse and longitudinal electron momentum components, for ionization of He by 3.6 MeV/amu Au^{53+} impact. The curves show a CDW-EIS calculation.

(the square of which yields the cross section) is expanded in powers of the interaction potential. For $\eta > 1$, it is not clear whether the expansion even converges. CDW-EIS, on the other hand, which accounts for higher-order contributions in the wavefunction rather than in the operator of the transition amplitude, is not directly affected by such convergence problems.

One might argue that the data of Figure 2.3 do not very convincingly demonstrate all the advantages of DDCS compared to total cross sections, except that they offer a more sensitive test of theory. The DDCS for large electron emission angles do not seem to exhibit more pronounced structures in the electron energy dependence than the projectile energy dependence of the total cross-sections. However, a remarkably sharp structure is present in the DDCS for electrons ejected in the projectile beam direction. Initially, a relatively broad "bump" was found for 300 keV p + He collisions for an ejection angle of 10° [23]. This observation prompted theoretical calculations which predicted a very sharp and tall structure, now known as the "cusp," for an ejection angle of $0°$ at an electron energy corresponding to an electron speed equal to the projectile speed [24, 25]. This prediction was experimentally confirmed by Crooks and Rudd [26].

A more recent example of a double differential electron spectrum showing a cusp, taken for 28.5 MeV $F^{9+} + H_2$ collisions [27], is plotted in Figure 2.4. Apart from the cusp at about 800 eV a second structure near 3 keV, known as the binary encounter peak (BEP) is also visible. The dotted line represents an FBA calculation by Brauner and Macek [28], the thin solid line a higher-order calculation by the same authors, the dashed and thick solid lines two different versions of the CDW-EIS model by Schultz and Reinhold [29], and the open squares a classical trajectory Monte Carlo (CTMC) simulation by the same authors. The BEP is reproduced by all calculations (including the FBA), demonstrating that it is due to a first-order mechanism. It can be understood in terms of a very close collision between the projectile and the electron, where the residual target ion is essentially completely passive. In the rest frame of the projectile, the electron approaches the former at the projectile speed v_0 (in the laboratory frame) and gets back-scattered without changing its speed. Therefore, the electron changes its momentum by $2mv_0$, which, after transforming back to the laboratory frame, corresponds to an ejected electron energy of $2mv_0^2$.

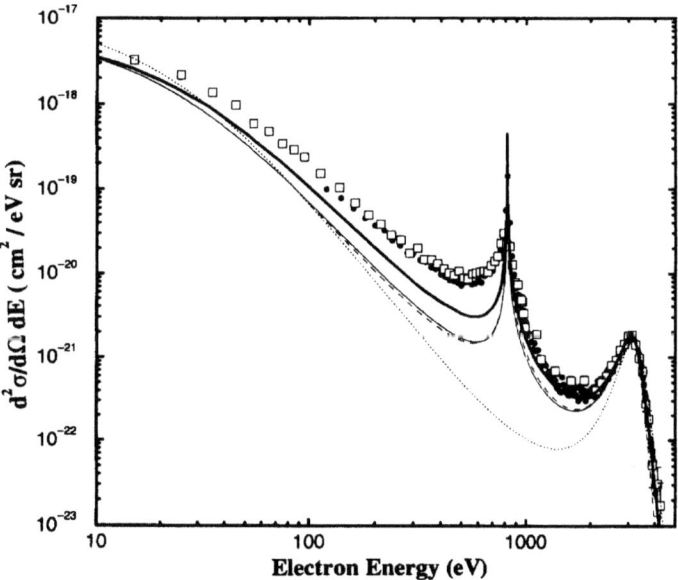

Figure 2.4: Double differential cross-sections, in ejected electron energy and electron solid angle, for ionization of H_2 by 28.5 MeV F^{9+} impact for an electron ejection angle of zero as a function of electron energy. The dotted curve represents the FBA model the other curves various versions of the CDW-EIS model. The open squares represent a classical approach based on a CTMC simulation.

On the other hand, the cusp is only predicted by the higher-order calculations and is completely missing in the FBA. This reinforces the interpretation that the cusp is due to a higher-order mechanism. Remarkably, the best agreement with the experimental data is obtained with the CTMC simulation, suggesting that the underlying ionization mechanism can to a large extent be understood within a classical picture. Since the cusp is reproduced by the CDW-EIS model, which does not include the NN interaction, it is clear that this mechanism must entail two (or more) interactions between the projectile and the electron. However, classically, two objects cannot undergo two subsequent collisions with each other, unless at least one of them is redirected by an interaction with a third body between these two collisions. Otherwise, the first two objects will continuously depart from each other after the first collision. In the case of ionization, the third body is the nucleus of the target atom, which can redirect either the projectile (NN interaction) or the electron after the first collision. Therefore, in a classical picture (like the CTMC simulation) two sequences of two-body interactions can lead to the cusp [30]. Both of them start with the primary interaction between the projectile and the electron lifting the latter to the continuum. In the first sequence, the projectile then elastically scatters off the target nucleus to be redirected towards the electron. We refer to this sequence as $V_{pe} - V_{NN} - V_{pe}$. In the second sequence, dubbed $V_{pe} - V_{Ne} - V_{pe}$, the ejected electron is "spun around" by the target nucleus to be redirected toward the projectile. The last step in both sequences is a second interaction (known as post-collision interaction or PCI) between the ejected electron and the outgoing projectile. Since this interaction is attractive, the electron is accelerated towards the projectile, both in speed and in direction. Only the second sequence is accounted for by the CDW-EIS model.

As a final note of this section, we point out that in spite of the large height of the cusp its contribution to the total cross section is negligible. The reason is that the total cross section is obtained from double integration over the electron energy and solid angle Ω_e. Since $d\Omega_e$ is given by $\sin\theta_e d\theta_e d\varphi_e$ and the cusp only occurs within a very narrow angular range around $\theta_e = 0$, it hardly contributes at all to the integral.

4 Differential scattered projectile spectra

In the previous section, we mentioned that effects due to the NN interaction may not be easily observable in double differential electron spectra. After all, in this interaction momentum is only exchanged between the two nuclei, but one might suspect that the electron momentum is unaffected. This argument should be handled cautiously in view of the discussion of the cusp in the previous section. There, we explained the cusp in terms of two interaction sequences, one of which containing the NN interaction. Although the NN interaction does not affect the electrons directly, in this sequence it is an essential preceding step to PCI. Therefore, the NN interaction can have a significant impact on the magnitude of the cusp. Nevertheless, it is not surprising that at angles larger than approximately 10°, where the cusp plays no role at all, the DDCS are basically unaffected by the NN interaction. Based on this reasoning, one would expect that either differential projectile or recoil-ion spectra would be more feasible to observe effects due to the NN interaction since it directly changes the momentum of both particles.

In Figure 2.5, ionization cross-sections differential in the projectile solid angle for 3 MeV p + He collisions are plotted as a function of the projectile scattering angle θ_p [31]. In the data, a distinct change of slope is observed at approximately 0.5 to 0.6 mrad. This feature can also be explained within a classical picture of the ionization process. In this picture, we initially treat the electron as quasi-free because the ionization potential is relatively small (24.6 eV). The ionization process can then be viewed as elastic scattering between the projectile and the electron in which at least 24.6 eV of kinetic energy is transferred to the electron. The residual target ion is regarded as a passive spectator. Applying momentum and kinetic energy conservation yields the following relation between the final projectile speed v_f and θ_p:

$$\cos \theta_p = [v_0^2(M - m) + v_f^2(M + m)]/(2Mv_0v_f) \qquad (2.1)$$

where v_0 is the initial projectile speed and m and M are the electron and projectile masses, respectively. This dependence of θ_p on v_f has a maximum, provided that $M > m$ (i.e., for any ionic projectile), which can be found by setting the derivative of $\cos \theta_p$ in v_f equal to zero, which yields the remarkably simple result:

$$\sin \theta_{\max} = m/M \qquad (2.2)$$

For proton–electron scattering one finds a maximum scattering angle (in the laboratory frame) of 0.54 mrad. This value exactly coincides with the angle where the change of slope in the data of Figure 2.5 occurs. The fact that the cross-sections do not rapidly drop to zero at this angle is due to two reasons: first, the application of momentum and kinetic energy conservation, yielding equation (2.1), neglects the initial motion of the electron in the target, which "smears out" the maximum scattering angle. Second, we treated the residual target ion as passive, i.e., we neglected any scattering of the projectile off the target ion. Since the target nucleus is more massive than the projectile there is no maximum scattering angle resulting from the NN interaction.

With this analysis, the data of Kamber et al. [31] can now be explained as follows: up to $\theta_p \approx 0.6$ mrad ionization is dominated by a first-order process only involving a single projectile–electron interaction. In this regime, higher-order processes are weak because of the small perturbation parameter ($\eta = 0.09$). At larger angles, the first-order process is only possible if the electron already has a large momentum in the initial target state. Therefore, with increasing angle ionization increasingly has to rely on higher-order mechanisms involving the NN interaction. This interpretation is also confirmed by theory. The curves in Fig. 2.5 show calculations by Fukuda et al. [32] and

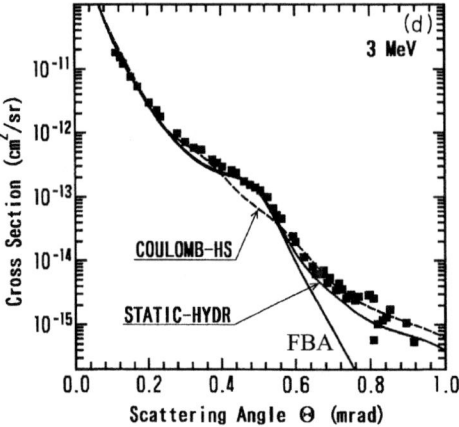

Figure 2.5: Single differential cross-section, in projectile solid angle, for ionization of He by 3 MeV proton impact as a function of scattering angle. The thick solid curve shows an FBA calculation, which does not include the NN interaction, while the other two curves represent higher-order calculations accounting for the NN interaction.

Salin [33], where the thick solid line represents the FBA and the dashed and thin solid lines variations of higher-order calculations accounting for the NN interaction. For $\theta_p < 0.4$ mrad the higher-order calculations and the FBA yield essentially the same results signifying that here ionization is dominated by the first-order mechanism. In contrast, for $\theta_p > 0.6$ mrad the FBA increasingly departs from the higher-order calculations, which very well describe the data, illustrating that higher-order contributions become increasingly important. However, in the integral over all solid angles, i. e., in the total cross-section, these contributions are negligible, as expected for this small value of η.

More detailed information about the role of the NN interaction in the ionization dynamics can be obtained when the projectile scattering angle dependence of the cross-sections is plotted for fixed ejected electron energies. These spectra represent double differential cross-sections in projectile solid angle (see footnote 1) and electron energy, to which we refer as $DDCS_p$ to distinguish them from double differential electron spectra. The first $DDCS_p$ were measured for 50 to 150 keV p + He collisions for electron energies ranging from about 5 to 100 eV [34, 35].

Later, $DDCS_p$ were measured for 75 keV p + H collisions [30, 36]. This is a particularly important collision system as it represents a pure three-body system (2 protons and one electron), i. e., the simplest system pertaining to the few-body problem. In Figure 2.6 data are shown for projectile energy losses of 30 and 40 eV (corresponding to electron energies of 16.4 and 26.4 eV). Calculations based on various models are shown, all of which conceptually account for higher-order contributions involving both multiple projectile-electron interactions and the NN interaction. The solid line represents an approach which can be regarded as a hybrid model, to which we refer

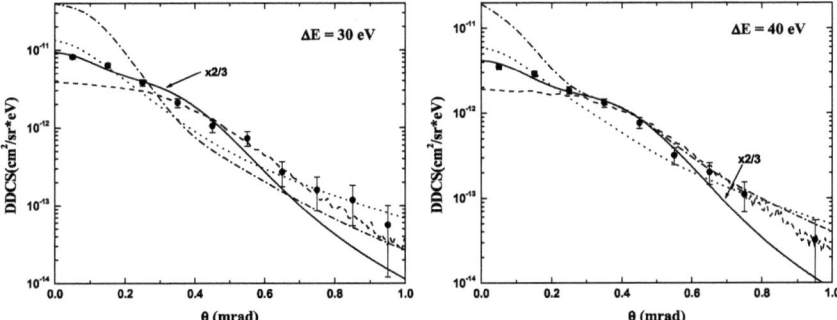

Figure 2.6: Double differential cross-sections, in projectile energy loss and solid angle, as a function of scattering angle for ionization of H by 75 keV proton impact. All curves represent higher-order models accounting for the NN interaction and multiple projectile-electron interactions.

as second-Born–Coulomb waves (SBA-C) model. Here, part of the higher-order contributions, those involving the NN interaction, is accounted for in the operator of the transition amplitude in terms of the second Born approximation. Higher-order contributions in the projectile-electron interaction, on the other hand, are treated, like in CDW-EIS, in the final state wavefunction of the collision system. Overall, the experimental data are best reproduced by the SBA-C results. However, there is a systematic discrepancy of a factor of 2/3 in absolute magnitude. This could partly be due to the systematic error in the experiment, which is estimated to be about 20 to 30 %. On the other hand, improved agreement in absolute magnitude with the measured data has been achieved by a very recent nonperturbative calculation [37].

In Figure 2.7, we show $DDCS_p$ for 3.6 MeV/amu Au^{53+} + He collisions [38] for various fixed ejected electron energies as a function of the momentum transfer component perpendicular to the initial beam axis $q_\perp$, which is unambiguously related to θ_p by $q_\perp = p_o \tan \theta_p$ (note that the largest $q_\perp$ for which data are shown corresponds to θ_p less than 1 µrad!). This is the same collision system for which the double differential electron spectra (Figure 2.3) are very well reproduced by the CDW-EIS model. The prediction of this approach for the $DDCS_p$ is shown as the dashed curves in Fig. 2.7. Based on the preceding discussion some discrepancies between the data and an approach not accounting for the NN interaction were to be expected. It is nevertheless quite remarkable that a calculation which works so well for the DDCS can go utterly awry for the $DDCS_p$. The solid lines show calculations which are based on a refinement of CDW-EIS to include the NN interaction [16, 38]. We refer to this model as CDW-EIS-NN. A drastic improvement in the agreement of the $q_\perp$-dependence of the cross-sections is quite apparent. Nevertheless, large discrepancies, especially in absolute magnitude, remain.

From the comparison between measured DDCS and $DDCS_p$ for various collision systems and the CDW-EIS and CDW-EIS-NN calculations, several conclusions can be drawn. First, the satisfactory agreement between theory and experiment in the DDCS

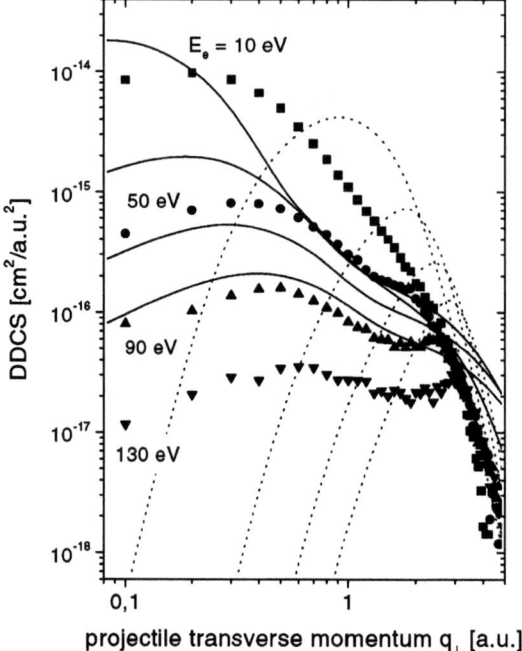

Figure 2.7: Double differential cross-sections for various fixed ejected electron energies as a function of the transverse momentum transfer (which is equivalent to the scattering angle) for ionization of He by 3.6 MeV/amu Au^{53+} impact. The dashed curves show CDW-EIS calculations (without NN interaction) and the solid curves CDW-EIS-NN calculations (accounting for the NN interaction).

even for very large η suggests that describing higher-order effects involving multiple projectile-electron interactions in the final-state wavefunction, as done in all CDW-EIS versions, is very adequate. Second, the dramatic discrepancies at very large η between measured $DDCS_p$ and CDW-EIS combined with the significantly improved agreement with CDW-EIS-NN shows that, not surprisingly, in this regime the NN interaction plays a crucial role in the few-body ionization dynamics. Finally, the remaining discrepancies with CDW-EIS-NN suggest that treating higher-order effects involving the NN interaction in the final-state wavefunction in perturbative approaches may not be sufficient. This is also supported by the observation that for 75 keV p + H collisions the SBA-C model, treating the NN interaction in the operator, yields better agreement with experiment than CDW-EIS-NN [30].

5 Fully differential cross-sections

As outlined in Chapter 1, the measurement of a fully differential cross-section (FDCS) requires to experimentally determine $3n - 4$ momentum components, where n is the number of unbound particles in the final state. For single target ionization $n = 3$ so that 5 momentum components need to be directly measured. Therefore, it is sufficient to detect and momentum analyze two of the final-state particles; the momentum of the undetected particle is obtained from momentum conservation. For electron impact ionization, most fully differential experiments detected the scattered projectile electron and the ejected target electron (for a review see [39]). In all fully differential ion impact ionization experiments, on the other hand, the momentum analyzed recoiling target ions were measured in coincidence either with the momentum analyzed scattered projectiles [e. g., 40–42] or with the momentum analyzed ejected electrons [e. g., 19, 43–48] (for a review, see [49]).

The high degree of dimensionality of FDCS also means that there are many different possibilities of presenting the data. One possible representation is illustrated in Figure 2.8, which sketches the kinematics of the collision. The blue arrows show the initial and final projectile momenta $\mathbf{p}_0$ and $\mathbf{p}_f$ as well as the momentum transfer vector $\mathbf{q}$ defined as $\mathbf{q} = \mathbf{p}_0 - \mathbf{p}_f$. The dotted blue line represents the so-called scattering plane spanned by $\mathbf{q}$ and $\mathbf{p}_0$ (and $\mathbf{p}_f$). The red arrow labeled $\mathbf{p_e}$ shows the ejected electron momentum, which together with $\mathbf{q}$ spans a second plane, represented by the solid red line, to which we refer as the electron emission plane. The azimuthal electron angle φ_e is the angle enclosed by the scattering and electron emission planes while the polar angle θ_e is measured relative to the initial projectile beam axis. In our presentation of the FDCS, we fix the magnitude of q (corresponding to a fixed θ_p) and the ejected electron energy (corresponding to a fixed p_e) and plot the FDCS in a polar, three-dimensional spectrum as a function of φ_e and θ_e. Since φ_e is measured relative to the scattering plane the azimuthal angle of $\mathbf{p}_f$ (or $\mathbf{q}$) is fixed at 0. Therefore, for each data point in such a three-dimensional plot 5 kinematic parameters are fixed, i. e., the data represent FDCS.

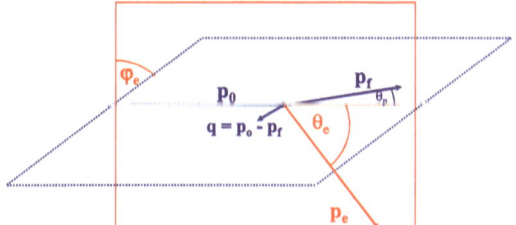

Figure 2.8: Illustration of the coordinate system and geometry in which fully differential cross-sections are presented. The momentum transfer $\mathbf{q}$ is defined as the difference between the initial and final projectile momenta $\mathbf{p}_0$ and $\mathbf{p}_f$ and $\mathbf{p_e}$ is the ejected electron momentum. The scattering plane (dotted blue line) is spanned by $\mathbf{p}_0$ and $\mathbf{q}$ and the electron emission plane (red line) by $\mathbf{p}_0$ and $\mathbf{p_e}$. The azimuthal electron angle φ_e is the angle between the scattering and electron emission planes and the polar angle θ_e is measured relative to $\mathbf{p}_0$. In the following figures, the FDCS are presented for fixed ejected electron energy and fixed q in a three-dimensional plot as a function of φ_e and θ_e.

An example of such a fully differential three-dimensional plot is shown in the upper left of Figure 2.9 for 100 MeV/amu C^{6+} + He collisions [43]. The electron energy was fixed at 6.5 eV and the momentum transfer at 0.75 a. u. (which corresponds to θ_p = 0.6 μrad). The blue and red lines (labeled I and II) represent the scattering plane and the plane perpendicular to it and to the projectile beam axis (perpendicular plane). For electrons ejected into the scattering plane, the angular dependence of the FDCS looks practically identical to what is routinely observed for fast electron impact [e. g., 50]: A large maximum, known as the binary peak, is found in the direction of **q**. A second, much smaller structure, dubbed the recoil peak, is observed in the direction of –**q**.

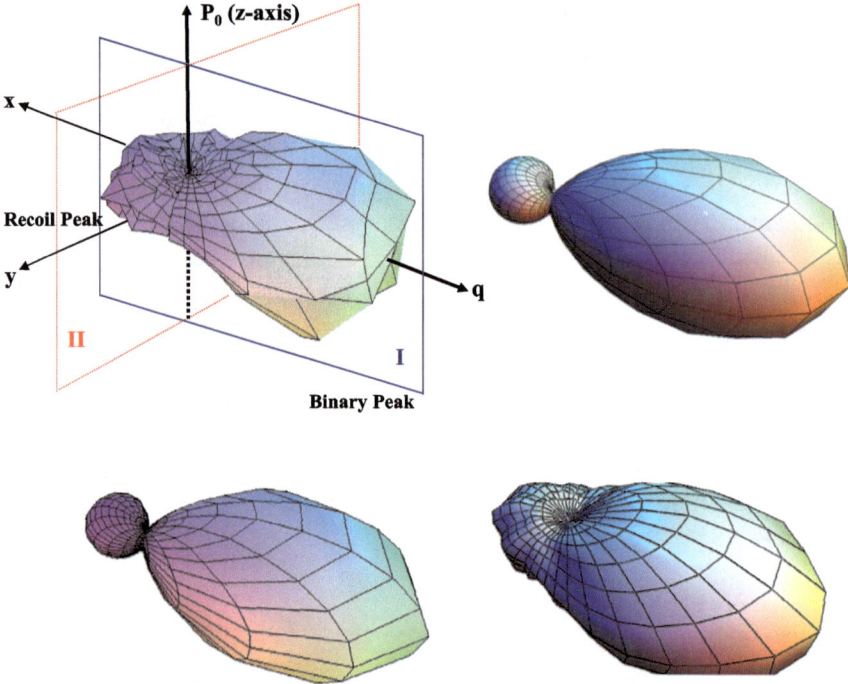

Figure 2.9: Three-dimensional, fully differential angular electron distribution for ionization of He by 100 MeV/amu C^{6+} impact for E_e = 6.5 eV and q = 0.75 a. u. Upper left panel, experiment; upper right panel, FBA calculation; lower left panel, 3DW calculation, lower right panel, FBA convoluted with classical elastic projectile-target nucleus scattering.

Both structures, the binary and the recoil peak, can qualitatively be explained within first-order models of ionization, like the FBA. The binary peak is due to a single interaction between the projectile and the electron. If the momentum distribution of the electron in the initial target state is neglected, momentum conservation then demands that the electron is emitted in the direction of **q**. The finite width of the binary peak is a reflection of the initial electron momentum distribution. The recoil peak is due to a two-step process: in the first step, the electron is lifted to the continuum by an interaction with the projectile, like in the binary peak, and in the second step the electron then gets back-scattered by the target nucleus so that the final electron momentum is pointing nearly in the direction of $-\mathbf{q}$. Note that the target nucleus–electron interaction is included to all orders in the electronic wavefunction so that the mechanism leading to the recoil peak is included even in the FBA although it is a double collision process. "First-order" refers to the projectile-target interaction, but not to any internal target interaction.

A calculation based on the FBA is shown in the upper right part of Figure 2.9. Here, the binary peak is centered exactly on $\mathbf{q}$ and the recoil peak exactly on $-\mathbf{q}$. Furthermore, the angular dependence in the xy plane (using the coordinate system superimposed on the experimental data plot) is exactly identical to the angular dependence in the xz plane (i. e., the scattering plane), which reflects that the FDCS are cylindrically symmetric about $\mathbf{q}$. Both features are indeed strict requirements of the FBA if the electron is initially in the 1s state of the target atom. Any departures from these properties were thought to be signatures of higher-order contributions. The plot in the lower left part of Figure 2.9 is a 3-body distorted wave calculation [51], which is conceptually similar to CDW-EIS-NN. Here, the centroids of the binary and recoil peak are nearly, but not exactly in the direction of $\mathbf{q}$ and $-\mathbf{q}$, respectively. Furthermore, the cylindrical symmetry about $\mathbf{q}$ is slightly broken in the 3-DW calculation. This suggest that in this model higher-order contributions are present, but relatively weak, as expected for collision system with such a small η.

In the experimental data, no statistically significant departure of the binary peak from the direction of $\mathbf{q}$ can be discerned. However, there is a substantial breaking of the cylindrical symmetry about $\mathbf{q}$. In the scattering plane (blue line), a deep minimum separating the binary from the recoil peak is visible, while in the xy plane (red line) there is only a shallow minimum at best. It is tempting to interpret this symmetry breaking as a manifestation of significant higher-order contributions for electrons ejected outside the scattering plane. Indeed, nonnegligible effects due to processes involving the NN interaction were found in theoretical calculations [e. g., 52–54]. However, fully quantum-mechanical calculations accounting for the NN interaction still showed severe discrepancies with the experimental data outside the scattering plane.

The treatment of the NN interaction in fully quantum-mechanical calculations is quite sophisticated (e. g., [53, 55, 56]) and it was difficult to understand how these models could go so severely wrong. Even more surprisingly classical or semiclassical calculations (like the one shown in the lower right panel of Fig. 2.9) yielded much better agreement with the experimental data outside the scattering plane (e. g., [54, 57–59]). A possible explanation emerges if one keep in mind that apart from neglecting all higher-order contributions the FBA makes another crucial approximation, with which other fully quantum-mechanical, but not (or only partly) the (semi-) classical calculations, are also afflicted: the projectiles are described by a de-localized wave (e. g. plane wave in the FBA), corresponding to a δ-function in momentum space. In other words, the projectiles are assumed to be coherent. Especially for fast heavy ions this is not necessarily a good approximation. It seems more realistic to account for large intrinsic momentum spreads for such projectiles by describing them in terms of a broad wave packet in momentum space, i. e., to treat the projectiles to be incoherent. It should be noted that this momentum spread does not reflect experimental resolution, but rather the quantum-mechanical momentum uncertainty associated with the Heisenberg relation. Several experimental and theoretical studies seem to support the explanation that the symmetry breaking of the FDCS about **q** is caused by such (de-) coherence effects (e. g., [58–66], for a review see [67]). Coherence effects are discussed in more detail in Chapter 4.

In Figure 2.10, the FDCS for ionization in 75 keV p + He collisions are plotted for electrons with an energy of 5.4 eV and momentum transfer fixed at 0.77 a. u. [40]. For this collision system, the perturbation parameter is much larger (η = 0.58) than for 100 MeV/amu C^{6+} + He and higher-order contributions should be considerably larger. For electron impact at similar η and for such a small ejected electron energy, the binary peak is usually shifted in the backward direction relative to the direction of **q** (e. g., [39]). This can be explained by higher-order mechanisms involving multiple projectile-electron scattering. Since the slow ejected electron lags behind the scattered projectile electron after the primary interaction, the repulsive force between both electrons leads to a momentum transfer in the second interaction which is pointing backward. With this argument, one would expect that for ion-impact, for which the projectile- electron interaction is attractive, the binary peak should be shifted in the forward direction relative to **q**. In the data of Figure 2.10, a shift in the backward direction is observed, instead.

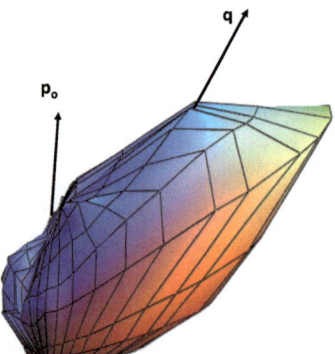

Figure 2.10: Three-dimensional, fully differential angular electron distribution for ionization of He by 75 keV proton impact for E_e = 5.4 eV and q = 0.77 a. u.

This backward shift of the binary peak has been explained within a classical picture, which is illustrated in Figure 2.11, in terms of a higher-order process involving the NN interaction [68]. In the primary interaction between the projectile (with an initial momentum $\mathbf{p_0}$) and the electron a momentum $\mathbf{q_e}$ is transferred to the electron, which is equal to the final electron momentum $\mathbf{p_e}$ if its initial momentum distribution in the target is neglected. This interaction leaves the projectile with an intermediate momentum of $\mathbf{p_m}$. If the projectile passes the target atom on the left, $\mathbf{q_e} = (\mathbf{p_0} - \mathbf{p_m})$ will have a transverse component to the left and, because the projectile loses energy in the collision, a longitudinal component in the forward direction (note that $\mathbf{q_e}$ is the negative of the corresponding change in projectile momentum $\Delta\mathbf{p} = \mathbf{p_m} - \mathbf{p_0}$). In the second step, the projectile then gets elastically scattered from the target nucleus. On average, the projectile approaches the electron to a closer distance than the nucleus. As a result, the transverse component of the momentum transferred to the recoiling nucleus $\mathbf{q_r}$ tends to be smaller than the transverse component of $\mathbf{q_e}$. At the same time, the longitudinal component of $\mathbf{q_r}$ is zero because essentially no energy is transferred in the elastic scattering between the two nuclei. Figure 2.11 shows that under these circumstances the total momentum transfer $\mathbf{q} = \mathbf{q_e} + \mathbf{q_r}$ is shifted forward relative to $\mathbf{q_e}$, or, in other words, $\mathbf{p_e}$ is shifted backward relative to $\mathbf{q}$, as observed in the experimental data.

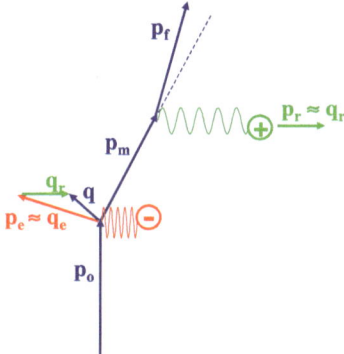

Figure 2.11: Classical view of a higher-order process involving the NN interaction which can lead to a backward shift of the binary peak relative to the direction of $\mathbf{q}$ (for details see text).

The shift of the binary peak was studied more systematically as a function of the transverse momentum transfer q_{tr} by Schulz et al. for 16 MeV O^{7+} + He collisions [47] for an ejected electron energy of 8 eV. The difference between the polar angle of $\mathbf{q}$ and the centroid polar angle of the binary peak $\Delta\theta = \theta_q - \theta_{BP}$ is plotted in Figure 2.12 as a function of q_{tr}. Positive $\Delta\theta$ correspond to a forward shift of the binary peak relative to $\mathbf{q}$ and negative $\Delta\theta$ to a backward shift. There are two areas of a forward shift in the data. At small q_{tr} of up to about 1 a. u. this shift is decreasing with increasing q_{tr} while for q_{tr} larger than 2 a. u. it is increasing with q_{tr}. Between 1 and 2 a. u., the binary peak is shifted in the backward direction relative to $\mathbf{q}$, as for 75 keV p + He collisions.

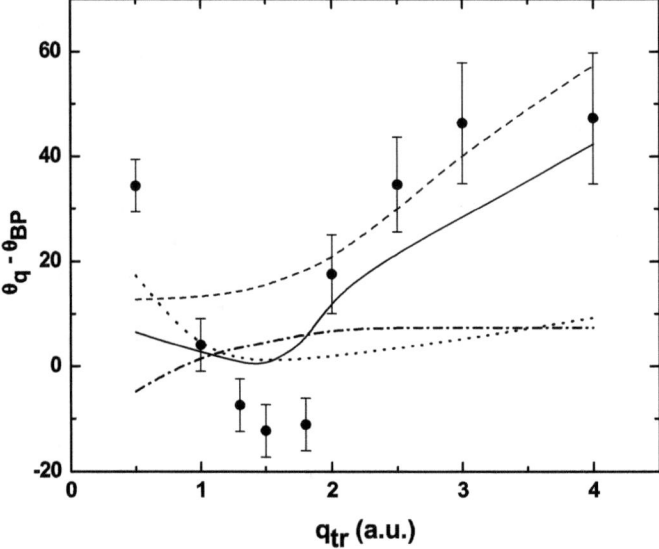

Figure 2.12: Difference angle between the direction of **q** and the centroid of the binary peak for electrons with an energy of 8 eV ejected into the scattering plane in 16 MeV O^{7+} + He collisions as a function of the transverse momentum transfer. For calculations, see text.

The data are qualitatively reproduced by a CDW-EIS-NN calculation, shown as the solid curve in Figure 2.12, although the minimum around q_{tr} = 1.5 a. u. does not result in a backward shift. The dashed curve represents a CDW-EIS calculation, which does not include the NN interaction. Here, the shift is also always in the forward direction, but increases monotonically with increasing q_{tr}. This suggests that it is the same higher-order mechanism with the interaction sequence $V_{pe} - V_{Ne} - V_{pe}$ leading to cusp electrons which is mostly responsible for the forward shift in the region $q_{tr} > 2$ a. u. It is tempting to conclude that in the region $q_{tr} < 1$ a. u. it is the sequence $V_{pe} - V_{NN} - V_{pe}$ leading to the forward shift since it is only found in theory if the NN interaction is included. However, the comparison between the dotted and dash-dotted curves in Figure 2.12 suggests that this may not be the case. Both models are based on the FBA, but the NN interaction is included in terms of an eikonal phase. The difference between both calculations is that in the dotted curve the ejected electron is described by a distorted wave and by a plane wave in the dash-dotted curve. This means that the interaction between the electron and the target ion after the ejection is included in the former, but not in the latter case. These calculations show that for the forward shift at small q_{tr} the second projectile electron interaction (i. e., PCI) is not needed, but the interaction between the electron in the continuum and the target ion is. Therefore, the sequence $V_{pe} - V_{NN} - V_{Ne}$ seems to be mostly responsible for the shift of the binary peak in this region. Furthermore, if the NN interaction is included in the calculation, but PCI and the interaction between the ejected electron and the target ion are not (dash-dotted curve), theory predicts a backward shift of the binary peak for small q_{tr}. Therefore, theory to some extent supports the explanation that the mechanism illustrated in Figure 2.11 could lead to the backward shift observed in the data for 75 keV p + He collisions (Figure 2.10).

The perturbation parameter for 16 MeV O^{7+} + He is $\eta = 1.1$, for which, as mentioned in the section on differential electron spectra, it is not clear whether the Born series even converges. The CDW-EIS-NN model, on the other hand, is capable of at least describing the data qualitatively. A much more severe test of theory was provided by measurements of FDCS for 3.6 MeV/amu Au^{53+} + He collisions, which correspond to an extreme perturbation of $\eta = 4.4$. These data are shown in Figure 2.13 for an ejected electron energy of 10 eV and momentum transfers of (from left to right) of 0.45, 0.65, and 1.0 a. u. [69, 70]. The dotted, dashed, and solid curves show calculations based on the FBA, CDW-EIS, and CDW-EIS-NN models, respectively. These data were recorded in the same experiment for which the double differential electron spectra (DDCS) are plotted in Figure 2.3. While the DDCS are at least qualitatively reproduced by the CDW-EIS model, even the CDW-EIS-NN calculations reveal dramatic discrepancies to the experimental data in the FDCS. The absolute magnitude deviates by as much as a factor of 50 and even the shape is not at all reproduced by theory.

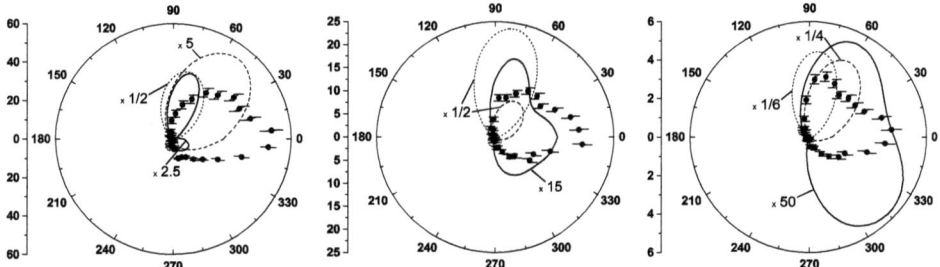

Figure 2.13: FDCS for ionization in 3.6 MeV/amu Au^{53+} + He collisions for electrons with an energy of 10 eV ejected into the scattering plane. The momentum transfer is fixed at, from left to right, 0.45 a. u., 0.65 a. u., and 1.0 a. u. The dotted, dashed, and solid curves show calculations based on the FBA, CDW-EIS, and CDW-EIS-NN models, respectively.

It is quite remarkable that a model which reproduces measured differential electron spectra so well is going completely awry in the FDCS. In fact, discrepancies become quite prominent as soon as cross-sections differential in the projectile solid angle are considered [38]. This suggests that theory has difficulties in describing reaction mechanisms in which the NN interaction is important. Those could partly stem from treating the projectiles completely coherently. Due to the much larger mass the de Broglie wave length of 3.6 MeV/amu Au^{53+} ions and, therefore, the coherence length, is even smaller than for 100 MeV/amu C^{6+} in spite of the significantly smaller speed. Indeed, a classical trajectory Monte Carlo (CTMC) calculation, which uses classical trajectories corresponding to a coherence length of zero, yielded improved agreement with the measured FDCS [71]. However, considerable discrepancies remained and it is unlikely that the poor agreement between fully quantum-mechanical calculations and experimental data is entirely due to coherence effects.

One disadvantage of the CDW-EIS-NN model, which is expected to cause problems at very large η, is that it is a perturbative approach. As mentioned earlier, this method should not be affected as severely by convergence problems as, e. g., the Born expansion. Nevertheless, the region of validity of CDW-EIS-NN is limited and was thought to be given by the requirement $\eta_c = Q_p/v_p^2 \ll 1$ (rather than $\eta = Q_p/v_p \ll 1$) [72]. For 3.6 MeV/amu Au^{53+} $\eta_c = 0.37$, which satisfies the above requirement only marginally. In this regime, one would expect improved agreement with non-perturbative models. One advantage of such approaches is that here the electronic basis usually consists of a large number of basis states, which are all coupled to each other (see Chapter 7). In contrast, perturbative models effectively only consider the initial and one final electronic state. As a result, reaction channels other than the one under investigation are ignored. When higher-order effects are important these reaction channels (like, e. g., capture) can divert a significant electron flux into basis states not corresponding to the process of interest. However, since in perturbative methods such states are not included, these reaction channels are erroneously counted as contributing to the process under investigation.

The FDCS for 3.6 MeV/amu Au^{53+} + He were calculated by the nonperturbative "coupled pseudostate" (CP) approach [73]. This is a very involved calculation with a basis set consisting of 165 states. It is therefore surprising that this model did not lead to substantially improved agreement with the experimental data compared to perturbative approaches. One approximation in this calculation is that the basis does not include any bound projectile states so that electron capture is not accounted for. Given the very high charge state of the projectiles one might suspect that this is a rather crude approximation. However, capture cross sections decrease very steeply with projectile speed (up to 12th power) and for 3.6 MeV/amu Au^{53+} + He collisions capture is actually negligible compared to ionization due to the large projectile speed.

The authors of [73] pointed out that in spite of the large basis set the calculations of the FDCS for 3.6 MeV/amu Au^{53+} + He may not be converged yet. However, they did feel very confident that their results for 3.6 MeV/amu Au^{24+} + He are well converged. For this smaller projectile charge state capture is even less important. Nevertheless, for this collision system, too, there are substantial discrepancies between experiment and theory. A calculation, describing the projectiles in terms of a wave packet with a width realistically reflecting the coherence length, could shed some light on the contributions of coherence effects to these discrepancies. If, on the other hand, such a calculation does not lead to substantially improved agreement with the data one has to conclude that the description of the effect of the NN interaction on the few-body dynamics still represents a major theoretical challenge in the regime of very large η.

6 Conclusions

Experimental techniques in ion-atom collision research have seen substantial developments over the last 20 to 30 years. With the advent of recoil-ion momentum spectroscopy [74, 75] kinematically complete experiments on ionization became feasible [4]. This, in turn, made possible experimental tests of theory with unprecedented sensitivity. The comparison between measured and calculated fully differential cross-sections revealed that the description of ion-impact ionization is much more challenging than for electron impact to a degree that was surprising. Depending on the perturbation parameter η of the collision, the difficulties stem from different sources.

Until recently, it was assumed that for small η the ionization dynamics is essentially understood, until this assumption was challenged by first kinematically complete experiments. The discrepancies between experiment and theory were blamed on not realistically considering the projectile coherence properties [45, 60] and this interpretation was supported by theoretical analysis (e. g., [64–66]). Indeed, if in the experiment a coherent projectile beam is prepared, very good agreement with theory is obtained [48]. Therefore, it does not seem overly optimistic to claim that ionization is understood for coherent projectiles with small η colliding with simple target atoms. Here, the collision dynamics is dominated by a first-order mechanism, i. e., the only important interaction is the one between the projectile and the active electron.

In contrast, at large η discrepancies between experiment and theory are severe (e. g., [70, 73]) and it seems unlikely that these can be entirely associated with coherence effects. Here, higher-order effects become very important. Differential electron spectra can be well described by sophisticated higher-order calculations. This suggests that higher-order contributions involving multiple projectile-electron interactions are satisfactorily understood even at very large η. However, as soon as cross-sections differential in projectile parameters are analyzed, which are sensitive to the NN interaction, the theoretical difficulties become quite apparent. Therefore, the description of the higher-order dynamics involving the NN interaction appears to be the largest remaining challenge.

In summary, the comparison between experimental and theoretical cross-sections shows that the fundamentally important few-body problem is far from being solved. In this context, it is important to remember that in atomic collision research one has the benefit that the underlying force, the electromagnetic interaction, is essentially understood. In nuclear collision research, the situation is even direr because our understanding of the underlying force (the strong interaction) is still rather incomplete so that testing the theoretical description of the few-body dynamics is much more difficult. Therefore, sustained efforts on studies of collisions, both atomic and nuclear, are crucially important.

Bibliography

[1]   Rudd ME, Kim Y-K, Madison DH, Gallagher JW. Rev. Mod. Phys. 1985;57(4):965.
[2]   Stolterfoht N, DuBois RD, Rivarola RD, Electron Emission in Heavy Ion-Atom Collisions. Heidelberg: Springer Verlag; 1997.
[3]   Rudd ME, Kim Y-K, Madison DH, Gay TJ. Rev. Mod. Phys. 1992;64(2):441.
[4]   Schulz M, Madison DH. International Journal of Modern Physics A. 2006;21:3649.
[5]   Keene JP. Philos. Mag. 1949;40:369.
[6]   De Heer FJ, Schutten J, Moustafa H. Physica. 1966;32:1766.
[7]   Afrosimov VV, Mamaev YuA, Panov MN, Uroshevich V. Zh. Tekh. Fiz. 1967;37:717.
[8]   Shah MB, Gilbody HB. J. Phys. B. 1981;14:2361.
[9]   Rudd ME, Dubois RD, Toburen LH, Ratcliffe CA, Goffe TV. Phys. Rev. A. 1983;28:3244.
[10] Kuyatt CE, Jorgensen T. Phys. Rev. 1963;130:1444.
[11] Rudd ME, Jorgensen T. Phys. Rev. 1963;131:666.
[12] Park JT, Aldag JE, George JM, Peacher JL, McGuire JH. Phys. Rev. A. 1977;15:508.
[13] McGovern M, Assafrão D, Mohallem JR, Whelan CT, Walters HRJ. Phys. Rev. A. 2009;79:042707.
[14] Knudsen H, Mikkelsen U, Paludan K, Kirsebom K, Møller SP, Uggerhøj E, Slevin J, Charlton M, Morenzoni E. Phys. Rev. Lett. 1995;74:4627.
[15] Tribedi LC, Richard P, Ling D, Wang YD, Lin CD, Moshammer R, Kerby GW, Gealy MW, Rudd ME. Phys. Rev. A. 1996;54:2154.
[16] Rodriguez VD, Barrachina RO. Phys. Rev. A. 1998;57:215.
[17] Sanchez MD, Cravero WR, Garibotti CR. Phys. Rev. A. 2000;61:062709.
[18] Foster M, Madison DH, Peacher JL, Schulz M, Jones S, Moshammer R, Ullrich J. J. Phys. B. 2004;37:1565.
[19] Voitkiv AB, Najjari B, Moshammer R, Schulz M, Ullrich J. J. Phys. B. 2004;37:L365.
[20] Pedlow RT, O'Rourke SFC, Crothers DSF. Phys. Rev. A. 2005;72:062719.
[21] Ciappina MF, Cravero WR, Schulz M, Moshammer R, Ullrich J. Phys. Rev. A. 2006;74:042702.
[22] Schmitt W, Moshammer R, O'Rourke FSC, Kollmus H, Sarkadi L, Mann R, Hagmann S, Olson RE, Ullrich J. Phys. Rev. Lett. 1998;81:4337.
[23] Rudd ME, Sautter CA, Bailey CL. Phys. Rev. 1966;151:20.
[24] Salin A. J. Phys. B. 8. Proc. Phys. Soc., London. 1969;2:631.
[25] Macek J. Phys. Rev. A. 1970;1:236.
[26] Crooks GB, Rudd ME. Phys. Rev. Lett. 1970;25:1599.
[27] Lee DH, Richard P, Zouros TJM, Sanders JM, Shinpaugh JL, Hidmi H. Phys. Rev. A. 1990;41:4816.
[28] Brauner M, Macek JH. Phys. Rev. A. 1992;46:2519.
[29] Schultz DR, Reinhold CO. Phys. Rev. A. 1994;50:2390.
[30] Schulz M, Laforge AC, Egodapitiya KN, Alexander JS, Hasan A, Ciappina MF, Roy AC, Dey R, Samolov A, Godunov AL. Phys. Rev. A. 2010;81:052705.
[31] Kamber EY, Cocke CL, Cheng S, Varghese SL. Phys. Rev. Lett. 1988;60:2026.
[32] Fukuda H, Shimamura I, Végh L, Watanabe T. Phys. Rev. A. 1991;44:1565.
[33] Salin A. J. Phys. B. 1989;22:3901.
[34] Vajnai T, Gaus AD, Brand JA, Htwe W, Madison DH, Olson RE, Peacher JL, Schulz M. Phys. Rev.

Lett. 1995;74:3588.

[35] Schulz M, Vajnai T, Gaus AD, Htwe W, Madison DH, Olson RE. Phys. Rev. A. 1996;54:2951.

[36] Laforge AC, Egodapitiya KN, Alexander JS, Hasan A, Ciappina MF, Khakoo MA, Schulz M. Phys. Rev. Lett. 2009;103:053201.

[37] Abdurakhmanov IB, Bailey JJ, Kadyrov AS, Bray I. Phys. Rev. A. 2018;97:032707.

[38] Moshammer R, Perumal AN, Schulz M, Rodriguez VD, Kollmus H, Mann R, Hagmann S, Ullrich J. Phys. Rev. Lett. 2001;87:223201.

[39] Ehrhardt H, Jung K, Knoth G, Schlemmer P. Z. Phys. 1986;D1:3.

[40] Maydanyuk NV, Hasan A, Foster M, Tooke B, Nanni E, Madison DH, Schulz M. Phys. Rev. Lett. 2005;94:243201.

[41] Hasan A, Sharma S, Arthanayaka TP, Lamichhane BR, Remolina J, Akula S, Madison DH, Schulz M. J. Phys. B. 2014;47:215201.

[42] Arthanayaka TP, Sharma S, Lamichhane BR, Hasan A, Remolina J, Gurung S, Schulz M. J. Phys. B. 2015;48:071001.

[43] Schulz M, Moshammer R, Fischer D, Kollmus H, Madison DH, Jones S, Ullrich J. Nature. 2003;422:48.

[44] Schulz M, Moshammer R, Perumal AN, Ullrich J. J. Phys. B. 2002;35:L161.

[45] Wang X, Schneider K, LaForge A, Kelkar A, Grieser M, Moshammer R, Ullrich J, Schulz M, Fischer D. J. Phys. B. 2012;45:211001.

[46] Hubele R, Laforge AC, Schulz M, Goullon J, Wang X, Najjari B, Ferreira N, Grieser M, de Jesus VLB, Moshammer R, Schneider K, Voitkiv AB, Fischer D. Phys. Rev. Lett. 2013;110:133201.

[47] Schulz M, Najjari B, Voitkiv AB, Schneider K, Wang X, Laforge AC, Hubele R, Goullon J, Ferreira N, Kelkar A, Grieser M, Moshammer R, Ullrich J, Fischer D. Phys. Rev. A. 2013;88:022704.

[48] Gassert H, Chuluunbaatar O, Waitz M, Trinter F, Kim H-K, Bauer T, Laucke A, Müller Ch, Voigtsberger J, Weller M, Rist J, Pitzer M, Zeller S, Jahnke T, Schmidt LPhH, Williams JB, Zaytsev SA, Bulychev AA, Kouzakov KA, Schmidt-Böcking H, Dörner R, Popov YuV, Schöffler MS. Phys. Rev. Lett. 2016;116:073201.

[49] Schulz M, Madison DH. International Journal of Modern Physics A. 2006;21:3649.

[50] Stefani G, Avaldi L, Camilloni R. J. Phys. B. 1990;23:L227.

[51] Madison DH, Schulz M, Jones S, Foster M, Moshammer R, Ullrich J. J. Phys. B. 2002;35:3297.

[52] Voitkiv AB, Najjari B, Ullrich J. J. Phys. B. 2003;36:2591.

[53] McGovern M, Whelan Colm T, Walters HRJ. Phys. Rev. A. 2010;82:032702.

[54] Olson RE, Fiol J. J. Phys. B. 2003;36:L365.

[55] Fischer DH, Madison D, Foster M, Schulz M, Moshammer R, Jones S, Ullrich J. Phys. Rev. Lett. 2003;91:253201.

[56] Colgan J, Pindzola MS, Robicheaux F, Ciappina MF. J. Phys. B. 2011;44:175205.

[57] Schulz M, Dürr M, Najjari B, Moshammer R, Ullrich J. Phys. Rev. A. 2007;76:032712.

[58] Jarai-Szabo F, Nagy L. Eur. Phys. J. D. 2015;69:4.

[59] Sarkadi L. Phys. Rev. A. 2018;97:042703.

[60] Egodapitiya KN, Sharma S, Hasan A, Laforge AC, Madison DH, Moshammer R, Schulz M. Phys. Rev. Lett. 2011;106:153202.

[61] Wang X, Schneider K, LaForge A, Kelkar A, Grieser M, Moshammer R, Ullrich J, Schulz M, Fischer D. J. Phys. B. 2012;45:211001.

[62] Schneider K, Schulz M, Wang X, Kelkar A, Grieser M, Krantz C, Ullrich J, Moshammer R, Fischer D. Phys. Rev. Lett. 2013;110:113201.

[63] Sharma S, Arthanayaka TP, Hasan A, Lamichhane BR, Remolina J, Smith A, Schulz M. Phys. Rev. A. 2014;90:052710.

[64] Arthanayaka T, Lamichhane BR, Hasan A, Gurung S, Remolina J, Borbély S, Járai-Szabó F, Nagy L, Schulz M. J. Phys. B. 2016;49:13LT02.

[65] Sarkadi L, Fabre I, Navarrete F, Barrachina RO. Phys. Rev. A. 2016;93:032702.
[66] Navarrete F, Ciappina MF, Sarkadi L, Barrachina RO. Nucl. Instrum. Meth. B. 2017;408:165.
[67] Schulz M. In: Arimondo E, Lin CC, Yelin S, editors. Advances in Atomic, Molecular, and Optical Physics, 66. Elsevier; 2017. p. 508–543.
[68] Schulz M, Hasan A, Maydanyuk NV, Foster M, Tooke B, Madison DH. Phys. Rev. A. 2006;73:062704.
[69] Fischer D, Moshammer R, Schulz M, Voitkiv A, Ullrich J. J. Phys. B. 2003;36:3555.
[70] Schulz M, Moshammer R, Perumal AN, Ullrich J. J. Phys. B. 2002;35:L161.
[71] Fremont F. Private communication (2018).
[72] Crothers DSF, McCann JF. J. Phys. B. 1983;16:3229.
[73] McGovern M, Assafrao D, Mohallem JR, Whelan CT, Walters HRJ. Phys. Rev. A. 2010;81:042704.
[74] Ullrich J, Moshammer R, Dorn A, Dörner R, Schmidt L, Schmidt-Böcking H. Rep. Prog. Phys. 2003;66:1463.
[75] Dörner R, Mergel V, Jagutzski O, Spielberger L, Ullrich J, Moshammer R, Schmidt-Böcking H. Phys. Rep. 2000;330:95.

Tom Kirchner

3 Collisions with antiprotons

1 Introduction

Antimatter particles have fascinated physicists (and nonphysicists) for many years and for good reasons. To begin with, their mere scarcity in the observable universe is one of the big mysteries of physics.[1] But there is more: Their interactions with matter particles may provide unique insights into the few-body problem (FBP) and give rise to a number of interesting applications; some of them chiefly based in science, some bordering on fiction, others crossing that border. Given that the FBP is a major theme of this book, this chapter on antiproton collisions will be mostly concerned with the question of how antiproton impact on atoms and molecules gives rise to phenomena and processes that add to our understanding of the FBP—or in cases in which we still lack a proper understanding, to our amazement about its complexity.

Antiprotons were first produced[2] and detected in 1955 at the Bevatron in Berkeley [1]. To carry out a meaningful atomic collision experiment, one needs an antiproton beam of well-defined kinetic energy in the keV to MeV regime. This is not at all what one finds in the production target of a high-energy accelerator such as the Bevatron: Those antiprotons have broad velocity distributions peaked at relativistic speeds. Accordingly, they need to be collected, slowed down and focused. The Low Energy Antiproton Ring (LEAR) was built at CERN[3] for this purpose.[4] It was in operation from 1982 to 1996 and a few years later replaced by the Antiproton Decelerator (AD), which is still on line. Most antiproton atomic collision experiments were carried out in the LEAR era by the CERN PS194 collaboration. Their work was followed up by the ASACUSA collaboration at the AD. A complete list of the cross-sections measured by both collaborations is provided in the Appendix of the review article [2]. The same

---

1 Presumably, matter and antimatter were created in equal amounts, but the antimatter seems to have disappeared almost completely shortly after the Big Bang leaving a strongly matter-dominated universe behind. The reasons for the excess of matter are not understood.

2 In a high-energy accelerator experiment via pair creation.

3 The acronym CERN is derived from the French *Conseil Européen pour la Recherche Nucléaire*. CERN's official full English name is European Organization for Nuclear Research.

4 Albeit mostly to support nuclear and particle physics research programs. Only a small fraction of the "tamed" antiprotons were allotted to atomic collision experiments.

Acknowledgement: This work was supported by the Natural Sciences and Engineering Research Council of Canada (NSERC).

Tom Kirchner, Department of Physics and Astronomy, York University, Toronto, Ontario M3J 1P3, Canada, e-mail: tomk@yorku.ca

https://doi.org/10.1515/9783110580297-003

work, and also previous review and original papers cited therein, give detailed accounts of the experimental techniques as well as most of the theoretical methods used to carry out cross-section calculations. These discussions will not be repeated here. Instead, we will look at a select subset of the results obtained and describe (in broad strokes) the physics they reflect.

The story will be told in largely chronological order as far as the experimental progress is concerned and is subdivided into five episodes. We will begin in Section 2 with a brief account of the first science question that was addressed by antiproton collision experiments: Is the previously observed factor-of-two difference in the total cross-section for double ionization of helium by equivelocity fast electrons and protons [3] a mass or a charge effect (or a combination of both)? Then, in Section 3, we will look at (mostly LEAR era) antiproton impact ionization cross-sections obtained at somewhat lower collision energies and demonstrate that they are well suited to test various theoretical approaches and learn about their validity and limitations. This sounds like a technical discussion and, indeed, some formalism and a few equations will be involved, but only to the extent as needed to spell out the ingredients of a given model and appreciate the physics it does or does not describe. We will then move down to even lower collision energies in Section 4, which became accessible experimentally with the commissioning of the AD. This is the regime of strong perturbations with projectile charge (magnitude)-to-speed ratios $\eta \geq 1$, which was mentioned in Chapter 1 of this book as a major motivation for pursuing antiproton collision studies in the first place. Indeed, we will see that interesting and quite unique physics happens here. Section 5 is concerned with potential applications of antiproton collisions. Naturally, we will focus on those which are based in science, but a little bit of fiction will be involved as well. The chapter ends with a look into the future of antiproton collision physics in Section 6.

2 Antiproton versus proton collisions at high impact energies: the role of the sign of the charge

Only two properties of antiprotons matter in the context of atomic collisions: mass and charge. In atomic units, the former is $m_{p^-} = 1836.15$ and the latter $q_{p^-} = -1$. A proton has the same mass and the opposite charge.[5] Given that the interactions in nonrelativistic collisions are purely Coulombic, changing the sign of the charge of the projectile corresponds to changing the sign of the coupling constant of the projectile-target interaction. The effects of this change can be appreciated in a perturbation-theory-based analysis in which the ionization amplitude is expanded in powers of the cou-

---

5 An electron has $m_{e^-} = 1$ and $q_{e^-} = -1$ and a positron $m_{e^+} = 1$ and $q_{e^+} = +1$.

pling constant. The cross-section σ is proportional to the squared modulus of the ionization amplitude. If the latter is expanded up to second order in the charge q, we have

$$\sigma(q) \approx A_2 q^2 + A_3 q^3 + A_4 q^4 \tag{3.1}$$

with coefficients A_2, A_3, A_4 which will be identical for proton and antiproton impact. Hence we obtain

$$\sigma_{p^+} - \sigma_{p^-} \approx 2A_3 \tag{3.2}$$

for the cross-section difference.

Let us consider proton and antiproton impact induced ionization of helium at $E = 1\,\text{MeV}$ collision energy. This corresponds to a projectile speed of 6.32 a. u. and a charge-magnitude-to-speed ratio of $\eta = 0.16$, i. e., low-order perturbation theory should be applicable (cf. Chapters 1, 2 and 5). For single ionization the measured total cross-sections are indistinguishable at this energy,[6] implying that a first-order calculation should be sufficient to describe this process. But the double ionization cross-section turned out to be higher for antiproton than for proton impact agreeing nicely with previously measured electron impact data [4]. This answered the question posed in Section 1: The difference is a charge effect and may be traced to a nonzero A_3 coefficient in equation (3.1). Put another way, first-order and second-order double ionization amplitudes appear to be of the same order of magnitude and interfere [5, 6].

One of the benefits of perturbation theory is that it is relatively straightforward to ascribe physical meaning to individual terms of an expansion. The second-order double ionization amplitude can be thought of as describing successive interactions of the projectile with each of the target electrons causing the removal of both. This has often been called the "two-step two" (TS2) mechanism [6]. In the first-order process, there is only one projectile-electron interaction, i. e., the projectile is directly responsible for ionizing only one of the two electrons. The second ionization step is thought of as being brought about in two different ways. In one of them the ionized electron strikes the other one on its way out, removing it as well. This is called the "two-step one" (TS1) or interception process [5, 6, 7]. Alternatively, the first electron may leave without interacting directly with the second one. If it leaves fast there is a sudden change in screening of the target nuclear charge and the second electron finds itself in a potential to which its state is not adapted. As a consequence, it may be shaken off. Both TS1 and shake-off processes have in common that they require the two electrons to interact in some way, or in more technical terms to be correlated with each

---

[6] As a note of caution one should mention that *absolute* ionization cross-sections were never measured for antiproton impact. Rather, it was found that the shape of the total single ionization cross-section as a function of impact energy was very similar for antiprotons and protons in the MeV regime. Given that $\eta \ll 1$ at such high energies, it was concluded that the cross-sections are the same.

other, in order to occur.[7] This is not the case for the TS2 process which is perfectly possible for two noninteracting electrons that move completely independently. These ideas can be understood more formally by realizing that the first-order amplitude is strictly zero if calculated in the framework of the independent particle model (IPM), while the second-order amplitude is not. The IPM is further discussed in the next section. Suffice it to say here that in the context of the FBP the IPM may be viewed as useful and convenient, but unexciting as it suggests that many-electron processes are in fact quite simple to understand. Conversely, any departure from independent particle dynamics may be taken as an indication of interesting few-body physics. It is then no surprise that the publication of the first experimental antiproton double ionization cross-sections was met with excitement and activity in theoretical quarters.

However, somewhat ironically an accurate second-order calculation that would confirm the argument given above in quantitative terms has never been reported. One problem is that one would need (quasi-) exact wave functions for two electrons in the continuum of He^{2+} to describe the final states, and another one the occurrence of a complete sum of intermediate (two-electron) states in the second-order term. Both issues can be dealt with only approximately and as a consequence of commonly used approximations the results obtained were never accurate or close enough to the experimental data to prove the argument unambiguously (see, e. g., [8]). Instead, largely numerical nonperturbative calculations have been carried out which support the experimental findings [9, 10]. Figure 3.1 shows the results of the first such calculation together with the experimental data for proton, antiproton, and electron impact. The plotted quantity is the ratio R of the double to single ionization cross-sections. Clearly, the sign of the projectile charge determines the outcome in the MeV regime with R being higher for the negatively charged projectiles than for proton impact.

The comparison between the antiproton and the electron impact data sheds light on the role of the other parameter: the projectile mass. It does not seem to play a role in very fast collisions, but gains importance toward lower energies where R for electron projectiles attains a maximum and then starts to fall off, while R for antiprotons continues to increase. A strict zero occurs for electron impact at an equivalent (anti) proton energy of 0.145 MeV. This corresponds to an electron energy of 79 eV, which is exactly the energy required to doubly ionize the helium atom. In other words, the zero marks the double ionization threshold. In the threshold region, the projectile mass matters because it limits the kinetic energy available to trigger an inelastic transition. Once the available energy is much higher than required for the process under investigation, its value, and hence the value of the projectile mass, lose importance. One

---

7 This is perhaps not obvious for the shake-off process, but can be explained by the fact that a correlated wave function is required for the description of the two-electron initial state to obtain a nonzero shake-off amplitude.

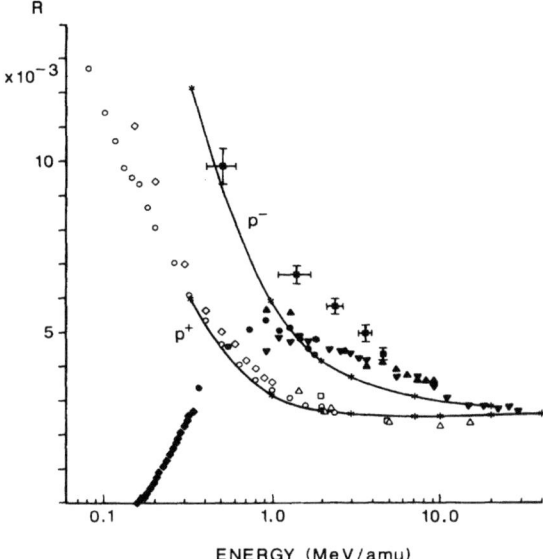

Figure 3.1: Ratio R of double to single ionization cross-sections for protons (p^+), antiprotons (p^-), and equivelocity electrons (e^-) colliding with He plotted as a function of impact energy. The full curves represent calculations from [9], scaled by a factor of 1.35 as described in that paper, while the symbols represent experimental data (properly cited in the same paper). The closed squares with error bars are for p^-, the other closed symbols for e^-, and the open symbols for p^+ impact. (Reproduced with permission from [9].)

may say that potential energy and, as a consequence, the projectile charge take over as the determining factors.

Finally, the data for all three projectiles appear to converge at $E \geq 20$ MeV,[8] implying that eventually the second-order TS2 mechanism becomes ineffective and double ionization proceeds as a pure first-order process—just like single ionization, but at much higher energies. This should not come as a surprise: A weaker process demands a smaller perturbation parameter η in order for first-order perturbation theory to be valid.

3 Antiproton collisions at intermediate impact energies: a test bench for theory

At projectile energies, $E < 1$ MeV differences in the antiproton versus proton impact cross-sections begin to show up in single ionization as well. This is demonstrated in

---

8 This is not obvious for the antiprotons due to the limited number of data points in Figure 3.1, but was confirmed by subsequent measurements at higher energies [11].

Figure 3.2 in which the cross-section difference (3.2), multiplied by a factor of 0.5, is plotted for the single ionization of helium. Two sets of data are shown corresponding to the fact that for the case of proton projectiles there are two fundamentally different ways to ionize the helium atom: The electron can either be released to the continuum or it can be captured by the projectile to form a neutral hydrogen atom. Obviously, the latter has no counterpart in antiproton collisions. If both processes are included, the cross-section difference (3.2) assumes large positive values at $E \leq 100$ keV. But even when capture is excluded proton and antiproton data differ from each other in most of the energy range shown, signaling that beyond-first-order calculations become a necessity below 1 MeV. Given the difficulties with explicit higher-order calculations, alternative approaches are advisable. One option is to still work within perturbation theory but use a distorted-wave approach (Chapter 5), another one to attack the problem fully nonperturbatively. This section focuses on the latter.

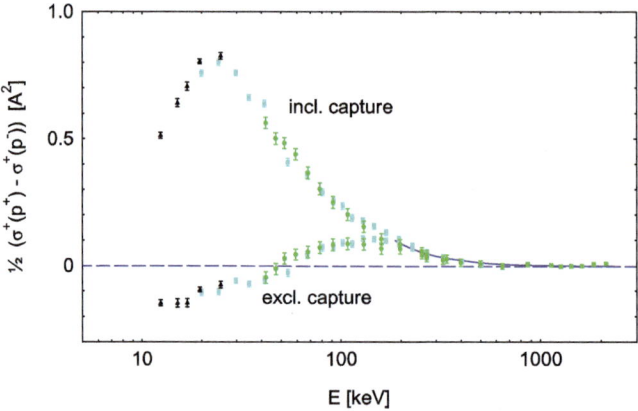

Figure 3.2: Cross-section difference (3.2) multiplied by 0.5 for single ionization of He by proton and antiproton impact plotted as a function of impact energy. The included data are cited in [2] and the full curve at high energies is not relevant for the present discussion. (Reproduced with permission from [2].)

In the following, a condensed account of what is described in more detail in [2] is provided in order to prepare for the discussion of a select set of *total* cross-section results. The starting point of most nonperturbative methods is the semiclassical approximation (SCA) of collision physics (for details, see Chapter 7). The SCA turns the stationary scattering problem of N electrons and, in the case of an atomic target, two heavy particles into a time-dependent Schrödinger equation (TDSE) for the electrons in the Coulomb field of a target nucleus at rest and a classically moving antiproton.[9]

---

9 For convenience, one usually chooses a reference frame whose origin coincides with the target nucleus.

Both heavy particles are no longer part of the many-body system of interest, but are merely there to provide an *external* potential for the electronic problem. Unless one is interested in very low collision energies ($E \lesssim 1\,\text{keV}$), one can approximate the motion of the antiproton by a straight-line trajectory $\mathbf{R}(t) = \mathbf{b} + \mathbf{v}_\text{p}t$ characterized by the impact parameter $\mathbf{b}$ and the constant projectile velocity $\mathbf{v}_\text{p}$ with $\mathbf{b} \cdot \mathbf{v}_\text{p} = 0$. This may seem crude, but the SCA with straight-line trajectories is known to be an essentially exact framework for total cross-section calculations at $E \gtrsim 1\,\text{keV}$ [12, 13]. The Hamiltonian in the electronic TDSE takes the form[10]

$$\hat{H}(t) = \hat{T} + \hat{V}_\text{ee} + \hat{V}_\text{ext}(t) \tag{3.3}$$

with the kinetic energy

$$\hat{T} = \sum_{j=1}^{N}\left(-\frac{1}{2}\nabla_j^2\right), \tag{3.4}$$

the electron–electron interaction

$$\hat{V}_\text{ee} = \sum_{i<j}^{N} \frac{1}{|\mathbf{r}_i - \mathbf{r}_j|}, \tag{3.5}$$

and the external Coulomb interaction with the target nucleus of charge number Z_T and with the negatively charged antiproton

$$\hat{V}_\text{ext}(t) = \sum_{j=1}^{N}\left(\frac{-Z_T}{r_j} + \frac{1}{|\mathbf{r}_j - \mathbf{R}(t)|}\right). \tag{3.6}$$

The Coulomb interaction between the antiproton and the target nucleus has been eliminated via a phase transformation, which leaves total cross-sections for inelastic processes unaffected [12].

The main difficulty in dealing with the Hamiltonian (3.3) is the two-body nature of the electron–electron interaction (3.5).[11] It prevents the TDSE from being separable in the electron coordinates and causes the much-debated electron correlation effects. For two-electron systems, i. e., collisions involving helium atoms, full solutions are feasible nevertheless and, accordingly, a number of explicitly-correlated calculations, beginning with the one shown in Figure 3.1, have been carried out over the years.

Beyond two electrons simplified approaches are needed. An obvious simplification is to replace the two-particle electron–electron interaction by effective one-body

---

10 We use atomic units, characterized by $\hbar = e = m_e = 4\pi\epsilon_0 = 1$, in this section.
11 Note that $\hat{V}_\text{ext}$ has one-body character.

operators in order to *make* the TDSE separable. This is the independent particle or independent electron model (IPM or IEM) [14]. It reduces the many-body TDSE to a set of effective single-particle equations and the electronic wave function to a simple (or better, an antisymmetrized) product state. Not surprisingly, the IEM has deficiencies when it comes to the comparison with experimental data (more on this further below). But surprisingly, one can formally prove that the representation of a many-electron system in terms of single-particle equations does not necessarily represent an approximation. The first version of such a proof for time-dependent problems was offered by Runge and Gross in 1984 [15] and the theory built on it is called time-dependent density functional theory (TDDFT). TDDFT is explained in depth and detail in the books [16, 17]. In the present context, it suffices to say that the central equations of TDDFT (the so-called Kohn–Sham equations) look like IEM equations, but they involve a potential that, in its exact form, includes electron correlation effects. The crux is that its exact form is not known, i. e., the (Kohn–Sham) potential has to be approximated or modeled in practical calculations. But there is more trouble. Even if the exact potential were available the only quantity that can be calculated directly from solving the Kohn–Sham equations is the (one-particle) electron density—a function in three-dimensional real space. The basic theorems of TDDFT assert that all observables of interest can be derived from the density, but except for a few special cases it is not known how to do this in practice without introducing further approximations.

In atomic collisions the observables of interest are cross-sections. Within the SCA, the total cross-section for a particular inelastic process is obtained by integrating the transition probability for that process over the impact parameter plane (i. e., the plane perpendicular to $\mathbf{v}_p$). Exploiting the cylindrical symmetry of the problem, this amounts to the integral

$$\sigma = 2\pi \int_0^\infty b P(b)\, \mathrm{d}b. \tag{3.7}$$

Let us consider the single and double ionization of helium. The corresponding probabilities can be readily expressed as integrals of the two-particle density over regions in space that can be associated with bound and unbound electrons. However, we would need to write the two-particle density in terms of the one-particle density to make those expressions directly workable. No one knows how to do this[12] except for the fictitious case of uncorrelated electrons, i. e., the IEM. In this case, the two-particle

---

12 One may wonder how this is possible at all given that the one-particle density n is an integral of the two-particle density ρ, i. e., n seems to contain much less information. A partial answer is that nobody said that the functional dependence of ρ on n will be local in time. In fact, there is good reason to believe that so-called memory effects play an important role [16, 17].

density is a simple product of one-particle densities (with appropriate normalization factor). Using the definition,

$$p = 1 - \frac{1}{2} \int_T n(\mathbf{r}, t) \, \mathrm{d}^3 r, \tag{3.8}$$

where n is the one-particle density and T some region in space that contains all target bound-state contributions to n, the probabilities for single and double ionization can be written as

$$P_1^{\mathrm{IEM}} = 2p(1 - p), \tag{3.9}$$

$$P_2^{\mathrm{IEM}} = p^2. \tag{3.10}$$

This is exactly what one would guess for two independent electrons with ionization probability p.

For the real case of correlated electrons, we can let us guide by equations (3.9) and (3.10) and cast the exact expressions into the form

$$P_1 = 2p(1 - p) - \mathcal{I}_c, \tag{3.11}$$

$$P_2 = p^2 + \frac{1}{2}\mathcal{I}_c, \tag{3.12}$$

where $\mathcal{I}_c$ is a so-called correlation integral that captures the departure of reality from the IEM [18]. Formally, equations (3.11) and (3.12) are still exact, but they are also still inexecutable since the exact expression for $\mathcal{I}_c$ involves the unknown two-particle density. So have we fooled ourselves? Not quite. The rewrite is useful as a starting point for approximations in which the "zeroth" order is the IEM ($\mathcal{I}_c = 0$) and higher orders include corrections to it.

Figure 3.3 shows results for the double ionization of helium by antiprotons obtained from using different TDDFT models [18]. Three variants clearly overestimate the experimental data at all but the highest impact energies. One of them, dubbed "OPM IEM," is a pure IEM calculation,[13] i. e., a calculation in which the correlation portion in the Kohn–Sham potential *and* the correlation integral in (3.11) and (3.12) are completely neglected. The other two include some correlation effects, but obviously not enough to make a significant difference (which is why we do not have to worry about the details which are provided in [18]).

The two curves without symbols are more interesting as they show different trends. One of them, dubbed "MCHF WB" includes a tiny correlation contribution to the Kohn–Sham potential[14] while the one from the work of Henkel et al. [19] does

---

[13] Using the optimized potential method (OPM) to generate the Kohn–Sham potential.

[14] MCHF stands for multiconfiguration Hartree–Fock. This is a correlated electronic structure method which was used to generate an accurate He ground-state wave function and, through an inversion procedure, a ground-state correlation potential to be included in the TDDFT calculation [18]. Dynamic correlation was not included in the Kohn–Sham potential.

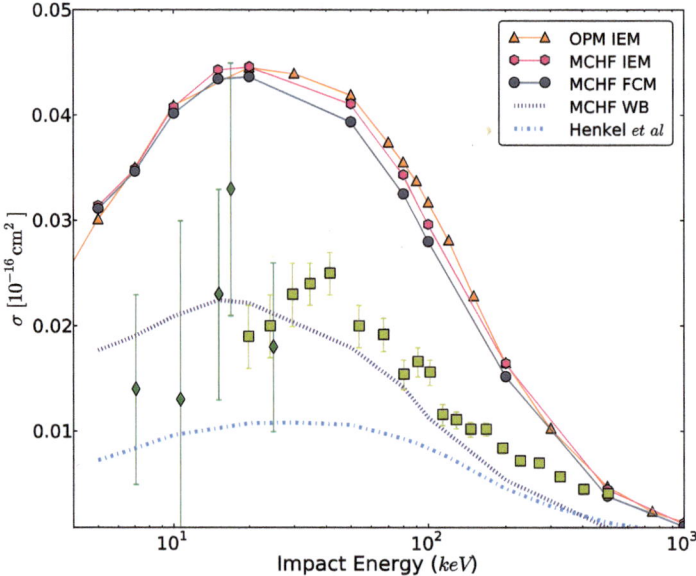

Figure 3.3: Total cross-section for double ionization of He by antiproton impact plotted as a function of impact energy. The symbols with error bars represent experimental data cited in [18]. The lines (with and without symbols) represent TDDFT calculations from [18] and Henkel et al. [19]. (Reproduced with permission from [18].)

not. The main difference of the two calculations however is how they approximate the correlation integral in (3.11) and (3.12). Both variants refer to a model proposed by Wilken and Bauer [20] in the context of laser-induced ionization, but only the MCHF WB calculation has that model for $\mathcal{I}_c$ properly implemented (hence the acronym WB). A vastly simplified version was used in [19]. The comparison with the experimental data demonstrates quite clearly that including a good model for $\mathcal{I}_c$ in the calculation does make a difference. One may speculate that the inclusion of a fully time-dependent correlation potential (if it would be available) would only serve to fine tune the results [18].

That some fine-tuning is indeed required is seen more clearly in Figure 3.4. It shows the p^--He double ionization cross-section once again, but this time the experimental data are compared to various full two-electron calculations plus the MCHF WB calculation of Figure 3.3. Clearly, the former do a better job than the TDDFT-based calculation at energies $E > 100$ keV. The fact that they do not agree perfectly among each other signals numerical issues, most likely associated with different finite basis sets that have been used by the various authors.[15]

---

15 The acronym TDCC used in Figure 3.4 stands for time-dependent close-coupling. TDCC methods are based on basis-set expansions of the electronic wave function and are perhaps the gold standard of ion-atom TDSE and TDDFT calculations (for details, see Chapter 7).

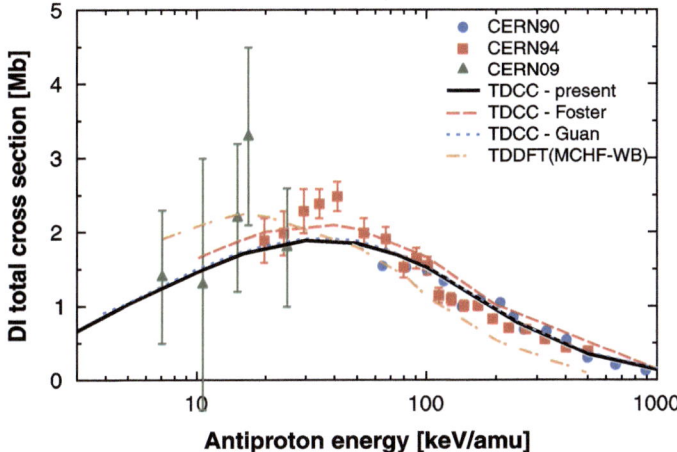

Figure 3.4: Total cross-section for double ionization of He by antiproton impact plotted as a function of impact energy. The symbols represent experimental data and the lines theoretical calculations cited in [21]. (Reproduced with permission from [21].)

Figure 3.5 shows the results of all of these plus a few more calculations for single ionization. We restrict ourselves here to summarizing the most important features of this triple-figure and refer the reader to [21] for a more detailed discussion. Panel (a) includes only full solutions of the two-electron TDSE which, if perfectly converged, should yield identical results. They do not, and interestingly show somewhat larger deviations among each other than for double ionization (cf. Figure 3.4). In panel (b), the two-electron calculation of [21] is compared with the MCHF WB calculation of [18] and with another two-electron calculation, dubbed CCC-MC (for convergent close-coupling multiconfiguration), that includes full correlation. The additional calculations shown in panel (c) account for electron correlation effects in a more restricted way or not at all.

All of these theoretical differences appear to matter mainly at energies below 100 keV. Given that most calculations are within the error bars of at least some of the experimental data points and since a smooth curve that would touch all points is difficult to picture, it is not clear which calculation is to be favored. The most prominent outlier is the calculation with the acronym CDW-EIS, which stands for continuum distorted-wave eikonal initial state. This is the only calculation displayed in Figure 3.5 which is based on perturbation theory. Most likely, the limitations of this framework cause the rapid decrease of the cross-section curve toward low impact energies which is in conflict with the "CERN08" data obtained at the AD.[16]

---

16 Curiously, two earlier (LEAR) experimental data points in the $10 < E < 20$ keV region appeared to support the rapid decrease, but they were deemed inaccurate and discarded when the AD measurements were published [22].

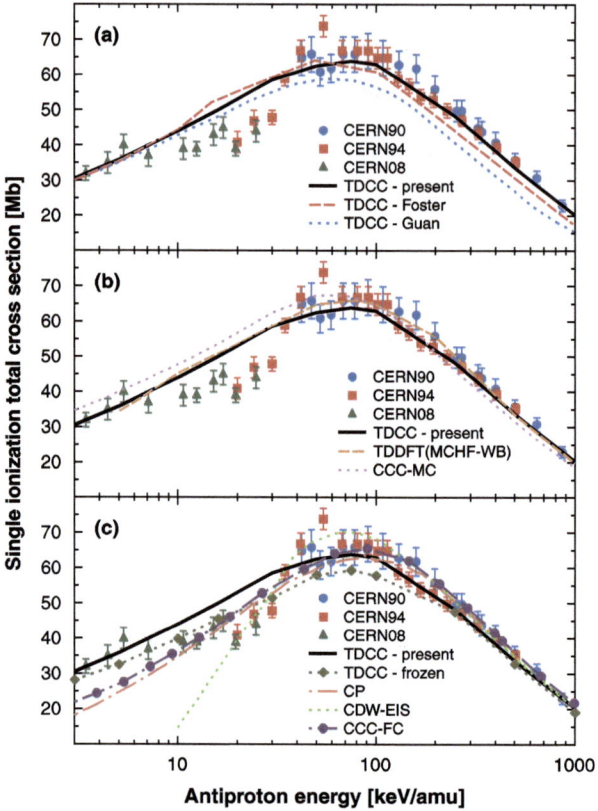

Figure 3.5: Total cross-section for single ionization of He by antiproton impact plotted as a function of impact energy. The symbols represent experimental data and the lines (with and without symbols) theoretical calculations cited in [21]. (Reproduced with permission from [21].)

Summarizing the discussion of p^--He collisions at energies $E < 1\,\mathrm{MeV}$, one may say that the total double ionization cross-section can only be properly described when electron correlation effects are taken into account. Full two-electron calculations do this in principle exactly and in practice sufficiently well if one measures them against the experimental data and their error bars. Furthermore, the double ionization cross-section serves as a playground for TDDFT, since one can learn about the relative importance of different TDDFT ingredients, namely the Kohn–Sham potential versus the correlation integral, by comparing results with the full calculations and the experimental data. This may help with the development of better models which potentially can then also be applied to other collision systems or even to other research areas such as laser-induced ionization.[17]

---

17 According to the basic theorems of TDDFT, the quantities in question are universal functionals of the density, i. e., their density dependence should be the same for different external interactions (3.6).

The situation is less clear for single ionization. Here, the spread of the results of a variety of nonperturbative calculations is mostly within experimental error bars, except for a possible structure in the experimental AD data in the 10–30 keV interval. The latter is not reproduced by any calculation and suggests that new measurements would be needed for a better understanding of the situation.

Helium is not the only atom that has been bombarded with antiprotons. Ionization experiments were also carried out for atomic hydrogen (see Figure 2.1 in Chapter 2), the heavier noble gases up to xenon, and a few molecules. Let us end this section with a look at the single ionization of neon and argon. The number of calculations carried out for these target atoms is significantly smaller than for helium. Figures 3.6 and 3.7, which are taken from the most recent paper on these collisions [23], show one nonperturbative calculation that includes correlation[18] ("CCC"), previous nonperturbative calculations on the IPM level ("IPM-BGM 1" and "IPM-BGM 2"),[19] a first-order perturbation theory calculation ("Born") and perturbative independent-electron CDW-EIS results [25].

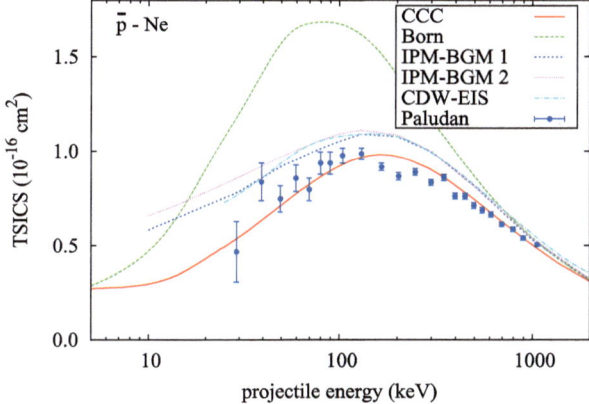

Figure 3.6: Total cross-section for single ionization of Ne by antiproton impact plotted as a function of impact energy. The symbols represent experimental data and the lines theoretical calculations cited in [23]. (Reproduced with permission from [23].)

The main purpose of the Born calculation in both figures is to show the (unsurprising, cf. Figure 3.2) limitations of first-order perturbation theory. The CDW-EIS results

---

18 In a so-called frozen-core framework in which a configuration-interaction wave function is built in such a way that only single excitations of the outer p-shell electrons are allowed. This is similar in spirit to the calculations included in Figure 3.5 dubbed "CCC-FC" and "TDCC-frozen."

19 BGM stands for basis generator method which is just another TDCC method in which a particularly constructed basis is used to represent the time-dependent orbitals [24].

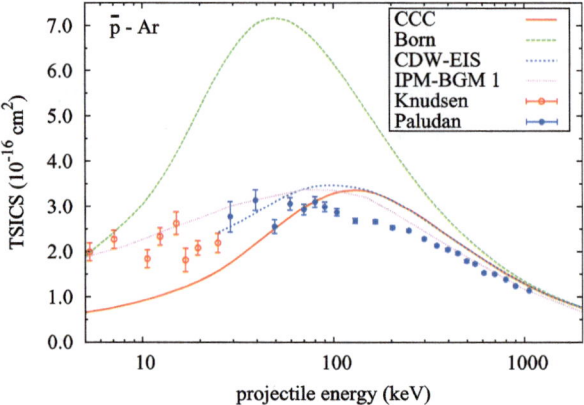

Figure 3.7: Total cross-section for single ionization of Ar by antiproton impact plotted as a function of impact energy. The symbols represent experimental data and the lines theoretical calculations cited in [23]. (Reproduced with permission from [23].)

demonstrate that the use of a distorted-wave formalism represents a substantial improvement. For neon in particular (Figure 3.6) the CDW-EIS results stack up quite well against the fully nonperturbative IPM-BGM calculations, whose two versions differ in the details of the effective potential substituted into the single-particle equations. The correlated CCC method yields smaller cross-sections. They are in excellent agreement with the experimental data for neon and suggest that correlation plays some role in this problem. For argon, however, things look somewhat different (Figure 3.7): Here, the CCC results appear to be too low at low energies (and too high at energies $E \gtrsim 100\,\text{keV}$), while the IPM-BGM 1 model shows good overall agreement with the experimental data (and the CDW-EIS results lie somewhere in between). This impression is in large parts due to the experimental data at $E < 25\,\text{keV}$, whereas the lowest energy point for neon is located at $E = 25\,\text{keV}$ and happens to have a large error bar. The low-energy argon data (depicted by red open circles in Figure 3.7) were obtained at CERN's AD [22], while all other data points for both atoms were measured in the LEAR era [26]. It would obviously be very interesting to see low-energy AD measurements for neon targets as well, or phrased more strongly: Without additional data points, we will not be able to say which level of theory is required for which level of accuracy of the total single ionization cross-section of heavy noble gases in this energy range.

As a final note, it should be mentioned that *multiple* ionization cross-sections were measured for neon and argon (and the heavier noble gases krypton and xenon) as well. Here, only IPM-level calculations (Born, CDW-EIS and BGM) are available for comparison. By way of disagreeing with the data, they indicate some role of electron correlation for the higher degrees of ionization, but perhaps a smaller one than for helium. At high energies, they suggest (very strongly) that a large fraction of multiple ionization events is due to post-collisional processes, such as Auger-electron emission after the direct ionization of an inner-shell electron. This adds another layer of complexity to the theoretical description of atomic collision processes—not just for antiproton impact. As was demonstrated in IPM-level works [25, 27, 28], post-collisional processes can be *modeled* in a relatively simple fashion. A first-principles based description is however outstanding.

4 Antiproton collisions at low impact energies: a unique situation

The discussion in the previous section indicates that measurements at the AD have added substantially to the antiproton collision story. They comprise the region of strong perturbations ($\eta \geq 1$) in which electron projectiles cannot ionize most atoms and positively-charged ions prefer to capture electrons instead of ionizing them into the continuum (cf. Figure 3.2). Perhaps the most interesting AD cross-section result available to date has been obtained for a molecular target—hydrogen [29].[20] The low-energy total single ionization cross-section was found to linearly decrease with decreasing impact velocity and turned out to be substantially smaller than predicted by a previously published full two-electron calculation [30]. The observation was interpreted as a molecular-structure-induced suppression of (nondissociative) ionization whereby an electron that is close to one nucleus in H_2 is pushed toward the other nucleus by the impinging antiproton and is able to remain bound. Time is of the essence for this mechanism to be effective, i. e., the lower the projectile energy, the more time the active electron has to distance itself from the antiproton and avoid removal [29].

The interpretation was later confirmed by two-electron CCC calculations [31]. Figure 3.8 shows that the results of these calculations are in good agreement with the experimental data at low projectile velocities, but somewhat underestimate them in faster collisions.[21] The figure also includes results for antiproton impact ionization of atomic hydrogen and helium to highlight the different velocity dependences of atomic and molecular ionization. More direct evidence for the proposed mechanism was provided by comparing snapshots of the one-electron density distributions in a p^--H_2 and a p^--H collision at $E = 1\,\text{keV}$ and the impact parameter $b = 1\,\text{a. u.}$ [31]. For the molecular case, the temporary "escape" of parts of the electron cloud to the more distant nucleus is clearly visible in those plots, while one sees how density is pushed directly into the continuum in the atomic case.

---

20 Actually, molecular deuterium was used, but this is unlikely to make any difference in the 2.4–10.6 keV impact energy interval studied in [29].

21 Before they agree again above $E = 90\,\text{keV}$ which cannot be seen in Figure 3.8, but was shown in a different plot in [31].

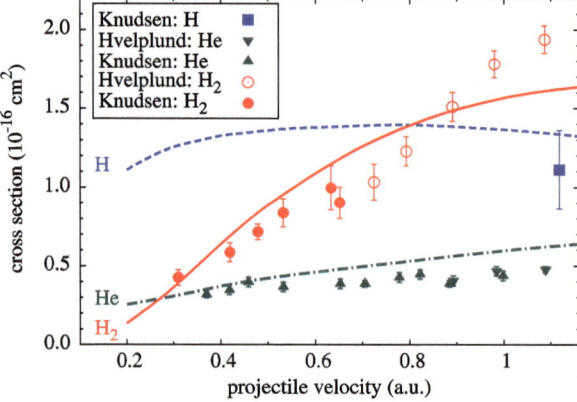

Figure 3.8: Total cross-sections for single ionization of H_2, He and H by antiproton impact plotted as functions of impact energy. The symbols represent experimental data cited and the lines CCC calculations reported in [31]. (Reproduced with permission from [31].)

A subsequent CCC calculation showed that structure-induced suppression of ionization is also at play in the one-electron p^--H_2^+ collision system [32], suggesting that electron correlation is not required for this effect to occur. Accordingly, it may well be that a TDDFT-based calculation at the IEM level is able to reproduce the two-electron CCC result for p^--H_2 at least qualitatively. In any case, the p^--H_2^+ calculations support the original conjecture that the mechanism may be a general feature in slow antiproton collisions with nonpolar molecules [29]. In polar molecules, stronger electrostatic forces may prevent the active electron to move away from the antiproton in an effective way. Clearly, further experimental and theoretical work is required to turn these ideas into established facts—or refute them.

At even lower projectile energies a new reaction channel opens up: antiproton capture, i.e., the formation of antiprotonic atoms. The ASACUSA collaboration at CERN has put significant effort into the creation, observation, and spectroscopy of antiprotonic helium[22][34, 35], in which one electron is replaced by an antiproton captured into a highly-excited state. Similarly, antiprotonic hydrogen, i.e., a bound antiproton–proton pair (also called protonium) was created following the interaction of antiprotons with H_2^+ ions [36]. Measurements of the formation cross-sections are not available,[23] but a number of theoretical calculations have been carried out. Figure 3.9 shows some of the results for antiproton collisions with helium in the 10–

---

[22] The main motivation for these experiments is the study of charge-parity-time reversal (CPT) invariance.

[23] A major difficulty is that the antiprotons are dominantly captured into highly excited states which in the case of antiprotonic helium may decay via Auger processes and evade observation unless they have sufficiently long lifetimes.

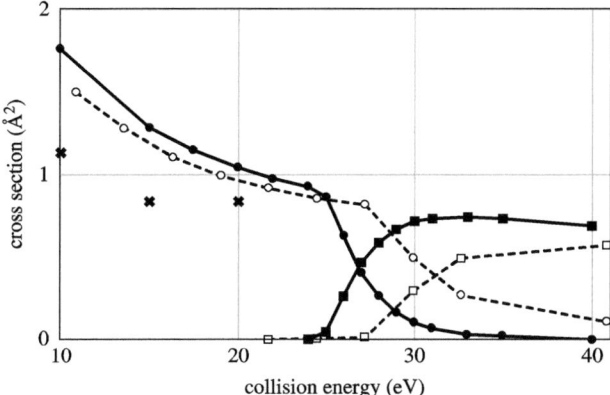

Figure 3.9: Total cross-sections for antiproton collisions with He resulting in single ionization (lines with open and filled squares) and in the formation of antiprotonic He (crosses and lines with open and filled circles) plotted as functions of impact energy. All theoretical results are cited in [33]. (Reproduced with permission from [33].)

40 eV energy range [33]. Target ionization is possible down to the threshold energy of $E = 24.6$ eV. Antiproton capture begins to play a role between 30–40 eV and takes over as the dominant and then the only inelastic channel toward lower energies. There is qualitative, but no quantitative agreement between the three theoretical models; a fully quantum-mechanical effective one-electron calculation (stars), a semiclassical effective one-electron calculation (full lines), and a quasi-classical two-electron, so-called fermion molecular dynamics, calculation (dashed lines). A fully quantum-mechanical two-electron calculation may be required to resolve the situation, but has not been reported yet in this energy interval.

5 Applications of antiproton collisions

Antiproton weapons are (thankfully) the subject of science fiction only and, likewise, antimatter fueled propulsion systems for space missions will not become a reality any time soon. But using antiproton beams to treat cancer may not be as crazy an idea as it sounds. In all three envisioned (or fantasized) applications, the proposition is to harness the energy that is released when antimatter and matter particles make contact and annihilate.

For antiproton cancer therapy the scenario is as follows: When a fast charged particle travels through matter, it loses kinetic energy, primarily because it tends to electronically excite and ionize the matter atoms or molecules along its path. In biological tissue, a number of secondary processes are initiated in this way, e. g., the released electrons may attack DNA molecules in ionizing collisions or via the formation of transient molecular resonances [37]. Damage to the DNA of a cancer cell may lead to the deactivation of the cell, i. e., the shutdown of its ability to create new cells via cell division. This is the desired effect of radiation therapy. Hadronic projectiles deposit most of their energy close to the end of their path giving rise to a prominent peak in a dose versus depth plot. The location of this so-called Bragg peak can be controlled via the beam energy. If it is adjusted such that it matches the position of a deep-seated tumor, chances are high that tumor cells are deactivated while the surrounding tissue is left unharmed.

The energy loss of fast antiprotons and protons in matter is similar (because their ionization and excitation cross-sections are), but there is a marked difference in what happens in the Bragg peak region: When an antiproton comes to rest it is captured to form an antiprotonic atom or molecule (see previous section) and annihilates on a nucleus eventually, thereby releasing 1.88 GeV of energy corresponding to twice the rest mass of the proton. Most of this energy is carried away (by pions), but 20–30 MeV are deposited close to the annihilation point and may contribute to the deactivation of cancer cells.

Figure 3.10 compares depth-dose curves for antiprotons, protons, and carbon ions in water obtained from a Monte-Carlo particle transport simulation. The doses are normalized to unity at the entrance point and the initial kinetic energies of the projectiles are adjusted such that the location of the Bragg peak coincides in all three cases. While the deposited dose is similar for all three particles in the plateau outside of the Bragg peak region, antiprotons are clearly more effective inside. This finding has been confirmed experimentally (see [38] and references therein). In addition, there is experimental evidence that the biological effects of the enhanced energy deposition are significant [39]. In an experiment conducted by the ACE collaboration at CERN living samples of a specific cancer cell type were irradiated by 50 MeV protons and antiprotons [40]. It was found that the ratio of surviving cells in the Bragg peak region compared to the plateau region closer to the target surface was significantly smaller for antiproton than for proton impact, signaling that tumor cell deactivation by antiprotons was more efficient. It was pointed out that this has two reasons: (i) the extra dose deposited in the Bragg peak due to antiproton annihilation, i. e., the approximate doubling of the peak seen in Figure 3.10, (ii) an enhanced biological effectiveness of the extra dose due to the fact that the additional 20–30 MeV are deposited in the form of short-range ions which are created by nuclear fragmentation processes following the annihilation event.[24]

Despite these favorable results, the road to the clinical implementation of antiproton therapy is a long and winding one. First of all, there is room for improvement in

---

24 On average 1–2 of the created pions will penetrate the remaining nucleus and cause its breakup [41].

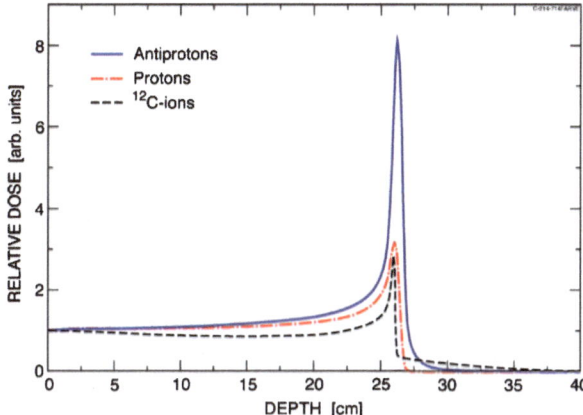

Figure 3.10: Normalized depth-dose profiles for antiprotons, protons, and carbon ions in a water target. Shown are the results of Monte-Carlo particle transport calculations cited in [38]. (Reproduced with permission from [38].)

the physics models and data that are used in the Monte-Carlo particle transport calculations. Accurate energy loss calculations for antiprotons in a number of targets have started to emerge [42, 43], but antiproton capture probabilities are modeled in relatively crude ways [41]. Moreover, other simulations using less ideal, and for clinical settings more realistic, spread-out antiproton beams suggest that the physical depth-dose distributions are in fact inferior to those of protons [44]. And then there is the looming question of the practicality of antiproton therapy: How expensive would it be to build and operate a dedicated facility, or would it be a more realistic alternative to deliver treatment parasitically at an existing or a future science laboratory? Clearly, much more work will be required to overcome the hurdles and answer these questions in order to make antiproton therapy a reality. But given the potential benefits, it is work worthwhile pursuing.

6 The future of antiproton collisions: differential cross-section measurements

All data shown so far in this chapter are total cross-sections. Yet, as discussed in Chapters 2 and 6 differential cross-sections provide additional, and sometimes deeper, insight into few-body collision dynamics, i. e., it seems desirable to undertake differential studies for antiproton impact as well. From an experimental viewpoint, this looks like a daunting task given that the antiproton beams used in the total cross-section measurements had rather low intensities and poor emittance properties. Nevertheless, one differential single ionization measurement was carried out some time ago for

945 keV antiproton impact on helium [45]. Not surprisingly at $\eta = 0.16$, it was found that the results were almost identical to proton impact data. One may perhaps characterize the work as an early heroic effort that should be followed up by measurements at lower projectile energies where first-order perturbation theory does not hold and the single ionization cross-sections for antiproton versus proton impact are expected to be different. Such measurements seem possible now thanks to the commissioning of the Extra Low ENergy Antiproton (ELENA) storage ring situated inside the AD at CERN. ELENA has been designed to decelerate the antiprotons down to $E = 100$ keV and deliver better-quality beams at this energy than previously available [35, 46]. An apparatus with similar or even superior capabilities may become available at the Facility for Antiproton and Ion Research (FAIR) in Darmstadt, Germany [47].

Meanwhile, the almost complete lack of experimental data has not prevented theorists from exploring differential antiproton collisions using perturbative and, more recently, also nonperturbative methods. If the interest is in cross-sections which are differential in the projectile deflection, it may appear necessary to abandon the SCA with straight-line trajectories and resort to a fully quantum-mechanical approach to the few-body problem of electrons, target nucleus, and antiproton. However, this is not necessarily true, since there exists a well-known and quite accurate method to convert semiclassical impact-parameter-dependent electronic transition amplitudes to projectile-angle-differential cross-sections [12]. The fact that this so-called eikonal approximation involves *transition amplitudes* rather than probabilities signals that phase information is important and indeed the internuclear interaction, which was eliminated from equation (3.3) by a phase transformation, reappears as a phase factor in the eikonal scattering amplitude.

The theoretically most-studied target is atomic hydrogen because of its relative simplicity as a pure one-electron problem.[25] Figure 3.11 presents an example from a recent work in which cross-section results from a first-order model ("FBA") and several fully nonperturbative calculations are compared with each other [48]. The plotted quantity is the fully differential cross-section (FDCS) for electron emission in the scattering plane at the impact energy $E = 200$ keV. The scattering angle and electron ejection energy are fixed at 0.2 mrad and 7 eV, respectively, and the FDCS is plotted as a function of the polar emission angle. An emission angle of zero degrees corresponds to electron emission in the direction of the momentum transfer vector.[26]

The most obvious feature of Figure 3.11 is the disagreement of the nonperturbative calculations with the FBA result. This implies that the FDCS for proton impact will be different, or in other words, that the antiproton FDCS is interesting. Indeed, a closer

---

25 Atomic hydrogen is however difficult to deal with in the laboratory and will most likely not be studied experimentally in the foreseeable future.

26 For a discussion of the general features of a single ionization FDCS, see Chapter 2. Note that the definition of the polar angle in Figure 2.8 of that chapter is different from that in Figure 3.11.

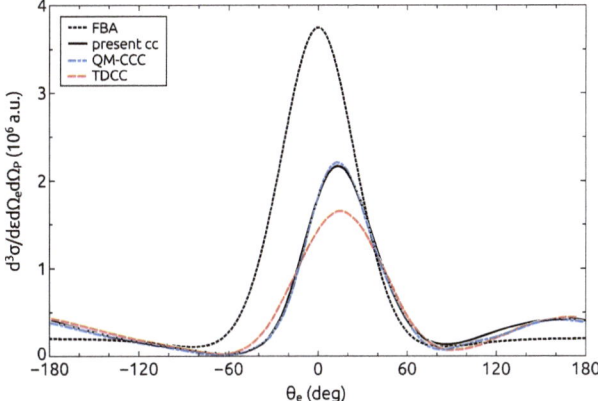

Figure 3.11: Fully differential cross-section for antiproton-impact ionization of atomic hydrogen at 200 keV in the scattering plane plotted as a function of the electron emission angle. The scattering angle of the projectile is 0.2 mrad and the ejected electron energy is 7 eV. The lines represent a first-order perturbation theory calculation (FBA) and three fully nonperturbative calculations cited in [48]. (Reproduced with permission from [48].)

analysis shows that in the FBA the dominant binary and the smaller recoil peaks have to show up at zero and 180 degrees, respectively, but are rotated away from the scattered antiproton in the nonperturbative calculations. The opposite happens in proton impact ionization. Similarly, the observed decrease of the binary peak and the increase of the recoil peak compared to the first-order result is reversed if the sign of the projectile charge is changed (see the discussion in [49] where the fully quantum-mechanical convergent close-coupling (QM-CCC) calculation included in Figure 3.11 was first presented). All three nonperturbative calculations of Figure 3.11 show these features, but only two out of three (QM-CCC and the one dubbed "present cc," which is based on the SCA and uses the eikonal approximation) agree with each other on a quantitative level. While this suggests that the deviating TDCC calculation of [50] (which also uses the eikonal approximation) is the deficient one, it would certainly be helpful to have experimental data available to confirm (or refute) this conjecture. As mentioned, one should not hope for fully-differential p^--H measurements any time soon, but one can imagine that similar features and discrepancies as those shown in Figure 3.11 may also occur for a more realistic system, such as p^--He.

Once differential measurements become feasible, it is easy to think of other interesting scenarios to explore and questions to ask. Here are a few examples: The first one concerns the suppressed ionization of molecules at low impact energies discussed in Section 4. What differential signatures might this mechanism exhibit? Secondly, it would be of considerable interest to study double ionization differentially. Some time ago an attempt was made to discuss on a differential level the effects of changing the sign of the projectile charge by comparing data for proton and electron impact on helium at high velocities, where mass effects should be unimportant [51]. Clearly, that discussion would be helped by corresponding antiproton measurements, just as the discussion of differences in total proton and electron double ionization cross-sections was settled with the publication of the first antiproton collision results in 1986. But what is more, with a machine like ELENA one may be able to initiate new discussions, e. g., by measuring differential double ionization at energies where antiprotons and electrons are *not* expected to yield similar results. In this region of large perturbation parameters η, one can expect unique features in antiproton impact double ionization spectra. Such measurements would challenge theorists to extend current nonperturbative approaches and in particular to come up with new ideas of how to represent the two-electron continuum. A number of relatively recent results suggest that the fine details of this representation are not important for total cross-section calculations [2], but this is likely different for differential cross-sections. Combined experimental and theoretical efforts in these areas will mark the beginning of a new chapter in the story of antiproton collisions. One can be confident that it will be an interesting one and that such future studies will add to our understanding of the few-body problem of dynamic systems.

Bibliography

[1] Chamberlain O, Segrè E, Wiegand C, Ypsilantis T. Observation of antiprotons. Phys. Rev. 1955;100:947–950.
[2] Kirchner T, Knudsen H. Current status of antiproton impact collisions: theoretical and experimental perspectives. J. Phys. B. 2011;44:122001.
[3] Puckett LJ, Martin DW. Analysis of recoil He$^+$ and He^{++} ions produced by fast protons in helium gas. Phys. Rev. A. 1970;1:1432.
[4] Andersen LH, Hvelplund P, Knudsen H, Møller SP, Elsener K, Rensfelt KG, Uggerhøj E. Single and double ionization of helium by fast antiproton and proton impact. Phys. Rev. Lett. 1986;57:2147.
[5] McGuire JH. Double ionization of helium by protons and electrons at high velocities. Phys. Rev. Lett. 1982;49:1153–1157.
[6] Knudsen H, Reading JF. Ionization of atoms by particle and antiparticle impact. Phys. Reports. 1992;212:107.
[7] Reading JF, Ford AL. The forced impulse method applied to the double ionisation of helium by collision with high-energy protons, antiprotons and alpha particles. J. Phys. B. 1987;20:3747.
[8] Nagy L, McGuire JH, Végh L, Sulik B, Stolterfoht N. Time ordering in atomic collisions. J. Phys. B. 1997;30:1939.
[9] Reading JF, Ford AL. Double ionization of helium by protons and antiprotons in the energy range 0.30 to 40 MeV. Phys. Rev. Lett. 1987;58:543.
[10] Foster M, Colgan J, Pindzola MS. Fully correlated electronic dynamics for antiproton impact ionization of helium. Phys. Rev. Lett. 2008;100:033201.
[11] Andersen LH, Hvelplund P, Knudsen H, Møller SP, Pedersen JOP, Tang Petersen S, Uggerhøj E, Elsener K, Morenzoni E. Further studies of double ionization of He, Ne, and Ar by fast and slow antiprotons. Phys. Rev. A. 1989;40:7366.

[12]   Bransden BH, McDowell MRC Charge exchange and the theory of ion-atom collisions. Oxford: Clarendon Press; 1992.
[13]   Abdurakhmanov IB, Kadyrov AS, Bray I, Stelbovics AT. Coupled-channel integral-equation approach to antiproton–hydrogen collisions. J. Phys. B. 2011;44:075204.
[14]   Ford AL, Reading JF Ion-atom and atom-atom collisions. In: Drake GWF, editor. Springer Handbook of Atomic, Molecular, and Optical Physics. New York: Springer; 2006. p. 753.
[15]   Runge E, Gross EKU. Density-functional theory for time-dependent systems. Phys. Rev. Lett. 1984;52:997–1000.
[16]   Ullrich CA Time-Dependent Density-Functional Theory: Concepts and Applications. Oxford: Oxford University Press; 2012.
[17]   Marques MAL, Maitra NT, Nogueira FMS, Gross EKU, Rubio A, editors. Fundamentals of Time-Dependent Density Functional Theory. Lecture Notes in Physics, 837. Berlin: Springer; 2012.
[18]   Baxter M, Kirchner T. Correlation in time-dependent density-functional-theory studies of antiproton-helium collisions. Phys. Rev. A. 2013;87:062507.
[19]   Henkel N, Keim M, Lüdde HJ, Kirchner T. Density-functional-theory investigation of antiproton-helium collisions. Phys. Rev. A. 2009;80:032704.
[20]   Wilken F, Bauer D. Adiabatic approximation of the correlation function in the density-functional treatment of ionization processes. Phys. Rev. Lett. 2006;97:203001.
[21]   Borbély S, Feist J, Tőkési K, Nagele S, Nagy L, Burgdörfer J. Ionization of helium by slow antiproton impact: Total and differential cross sections. Phys. Rev. A. 2014;90:052706.
[22]   Knudsen H, Kristiansen H-PE, Thomsen HD, Uggerhøj UI, Ichioka T, Møller SP, Hunniford CA, McCullough RW, Charlton M, Kuroda N, Nagata Y, Torii HA, Yamazaki Y, Imao H, Andersen HH, Tőkési K. Ionization of helium and argon by very slow antiproton impact. Phys. Rev. Lett. 2008;101:043201.
[23]   Abdurakhmanov IB, Kadyrov AS, Fursa DV, Avazbaev SK, Bailey JJ, Bray I. Antiproton-impact ionization of Ne, Ar, Kr, Xe, and H_2O. Phys. Rev. A. 2015;91:022712.
[24]   Kroneisen OJ, Lüdde HJ, Kirchner T, Dreizler RM. The basis generator method: optimized dynamical representation of the solution of time-dependent quantum problems. J. Phys. A. 1999;32:2141.
[25]   Montanari CC, Miraglia JE. Antiproton, proton and electron impact multiple ionization of rare gases. J. Phys. B. 2012;45:105201.
[26]   Paludan K, Bluhme H, Knudsen H, Mikkelsen U, Møller SP, Uggerhøj E, Morenzoni E. Single, double and triple ionization of Ne, Ar, Kr, and Xe by 30–1000 keV antiproton impact. J. Phys. B. 1997;30:3951.
[27]   Spranger T, Kirchner T. Auger-like processes in multiple ionization of noble gas atoms by protons. J. Phys. B. 2004;37:4159.
[28]   Montanari CC, Montenegro EC, Miraglia JE. CDW-EIS calculations for multiple ionization of Ne, Ar, Kr and Xe by the impact of H^+ and He^+, including post-collisional electron emission. J. Phys. B. 2010;43:165201.
[29]   Knudsen H, Torii HA, Charlton M, Enomoto Y, Georgescu I, Hunniford CA, Kim CH, Kanai Y, Kristiansen H-PE, Kuroda N, Lund MD, McCullough RW, Tőkési K, Uggerhøj UI, Yamazaki Y. Target structure induced suppression of the ionization cross section for very low energy antiproton-hydrogen collisions. Phys. Rev. Lett. 2010;105:213201.
[30]   Lühr A, Saenz A. Full two-electron calculations of antiproton collisions with molecular hydrogen. Phys. Rev. A. 2010;81:010701(R).
[31]   Abdurakhmanov IB, Kadyrov AS, Fursa DV, Bray I. Target structure-induced suppression of the ionization cross section for low-energy antiproton-molecular hydrogen collisions: Theoretical confirmation. Phys. Rev. Lett. 2013;111:173201.

[32] Abdurakhmanov IB, Kadyrov AS, Fursa DV, Avazbaev SK, Bray I. Close-coupling approach to antiproton-impact breakup of molecular hydrogen. Phys. Rev. A. 2014;89:042706.

[33] Jonsell S. Collisions involving antiprotons and antihydrogen: an overview. Phil. Trans. R. Soc. A. 2018;376:20170271.

[34] Hayano RS, Hori M, Horváth D, Widmann E. Antiprotonic helium and CPT invariance. Rep. Prog. Phys. 2001;70:1995.

[35] Hori M, Walz J. Physics at CERN's Antiproton Decelerator. Prog. Part. Nucl. Phys. 2013;72:206.

[36] Zurlo N, Amoretti M, Amsler C, Bonomi G, Carraro C, Cesar CL, Charlton M, Doser M, Fontana A, Funakoshi R, Genova P, Hayano RS, Jørgensen LV, Kellerbauer A, Lagomarsino V, Landua R, Lodi Rizzini E, Macrì M, Madsen N, Manuzio G, Mitchard D, Montagna P, Posada LG, Pruys H, Regenfus C, Rotondi A, Testera G, Van der Werf DP, Variola A, Venturelli L, Yamazaki Y. Evidence for the production of slow antiprotonic hydrogen in vacuum. Phys. Rev. Lett. 2006;97:153401.

[37] Boudaïffa B, Cloutier P, Hunting D, Huels MA, Sanche L. Resonant formation of DNA strand breaks by low-energy (3 to 20 eV) electrons. Science. 2000;287:1658.

[38] Knudsen HV, Holzscheiter MH, Bassler N, Alsner J, Beyer G, DeMarco JJ, Doser M, Hajdukovic D, Hartley O, Iwamoto KS, Jäkel O, Kovacevic S, Møller SP, Overgaard J, Petersen JB, Ratib O, Solbergj TD, Vranjes S, Wouters BG. Antiproton therapy. Nucl. Instrum. Methods Phys. Res. Sect. B. 2008;266:530.

[39] Holzscheiter MH, Alsner J, Bassler N, Boll R, Caccia M, Knudsen H, Maggiore C, Petersen JB, Sellner S, Straße T, Singers Sørensen B, Overgaard J. The relative biological effectiveness of antiprotons. Radiother. Oncol. 2016;121:453.

[40] Holzscheiter MH, Bassler N, Agazaryan N, Beyer G, Blackmore E, DeMarco JJ, Doser M, Durand RE, Hartley O, Kiwamoto KS, Knudsen HV, Landua R, Maggiore C, McBride WH, Pape Møller S, Petersen J, Skarsgard LD, Smathers JB, Solberg TD, Uggerhøj UI, Vranjes S, Withers HR, Wong M, Wouters BG. The biological effectiveness of antiproton irradiation. Radiother. Oncol. 2006;81:233.

[41] Taasti VT, Knudsen H, Holzscheiter MH, Sobolevsky N, Thomsen B, Bassler N. Antiproton annihilation physics in Monte Carlo particle transport code SHIELD-HIT12A. Nucl. Instrum. Methods Phys. Res. Sect. B. 2015;347:65.

[42] Bailey JJ, Kadyrov AS, Abdurakhmanov IB, Fursa DV, Bray I. Antiproton stopping power data for radiation therapy simulations. Phys. Medica. 2016;32:1827.

[43] Montanari CC, Miraglia JE. Low- and intermediate-energy stopping power of protons and antiprotons in solid targets. Phys. Rev. A. 2017;96:012707.

[44] Paganetti H, Goitein M, Parodi K. Spread-out antiproton beams deliver poor physical dose distributions for radiation therapy. Radiother. Oncol. 2010;95:79.

[45] Khayyat Kh, Weber T, Dörner R, Achler M, Mergel V, Spielberger L, Jagutzki O, Meyer U, Ullrich J, Moshammer R, Schmitt W, Knudsen H, Mikkelsen U, Aggerholm P, Uggerhøj E, Møller SP, Rodríguez VD, O'Rourke SFC, Olson RE, Fainstein PD, McGuire JH, Schmidt-Böcking H. Differential cross sections in antiproton- and proton-helium collisions. J. Phys. B. 1999;32:L73.

[46] Oelert W, Gabrielse G, Hayano R, Holzscheiter M, Hangst J, Lichard P ELENA: An upgrade to the Antiproton Decelerator. CERN-SPSC-2009-026, SPSC-P-338, 2009.

[47] Welsch CP, Papash AI, Gorda O, Harasimowicz J, Karamyshev O, Karamysheva G, Newton D, Panniello M, Putignano M, Siggel-King MRF, Smirnov A. Ultra-low energy storage ring at FLAIR. Hyperfine Interact. 2012;213:205.

[48] Bondarev AI, Kozhedub YS, Tupitsyn II, Shabaev VM, Plunien G, Stöhlker Th. Relativistic calculations of differential ionization cross sections: Application to antiproton-hydrogen collisions. Phys. Rev. A. 2017;95:052709.

[49] Abdurakhmanov IB, Kadyrov AS, Bray I, Stelbovics AT. Differential ionization in antiproton–hydrogen collisions within the convergent-close-coupling approach. J. Phys. B.

2011;44:165203.

[50] Ciappina MF, Lee T-G, Pindzola MS, Colgan J. Nucleus-nucleus effects in differential cross sections for antiproton-impact ionization of H atoms. Phys. Rev. A. 2013;88:042714.

[51] Fischer D, Moshammer R, Dorn A, Crespo López-Urrutia JR, Feuerstein B, Höhr C, Schröter CD, Hagmann S, Kollmus H, Mann R, Bapat B, Ullrich J. Projectile-charge sign dependence of four-particle dynamics in helium double ionization. Phys. Rev. Lett. 2003;90:243201.

R. O. Barrachina, F. Navarrete, M. F. Ciappina, and Michael Schulz

4 Coherence and contextuality in scattering experiments

1 Introduction

At a first sight, the description of a scattering experiment seems to be easy and unambiguous. A projectile, as for instance an electron or an ion, flies out of an accelerator with a given momentum **p** toward an atom or molecule, assumed to be at rest in
the collision chamber, and interacts with it. This interaction might lead to different
outcomes. For instance, the target might reach an excited bound state, or even loss
an electron. These are only two examples of a myriad of options, which are usually
called "channels." Thus, we can talk about the elastic channel, the excitation channel, the ionization channel, etc. Finally, one or some of the products of the collision
are collected in order to investigate one of these channels.

A closer look reveals that the description of any of these experiments is not quite
as straightforward. More specifically, one major complication is that the Schrödinger
equation is not analytically solvable for more than two mutually interacting particles,
even when the underlying forces are precisely known. This is known as the few-body
problem (FBP), which is formulated in the Introduction to this book and theoretically
discussed in other chapters. The FBP, in turn, is afflicted with an additional major
complication which is usually overlooked. Addressing this drawback of the standard
descriptions of the FBP is the main objective of the present chapter.

Both the initial preparation of the projectiles and the final detection of the outgoing particles occur at macroscopic distances and, therefore, it is assumed that they
cannot have any effect on the scattering event, which occurs in a region of atomic
dimensions. In other words, we might say that a scattering experiment is independent of its context. Furthermore, since both the beam of projectiles and the target gas
can be described by pure quantum states, we are dealing with a purely coherent process.

R. O. Barrachina, Centro Atómico Bariloche and Instituto Balseiro, Comisión Nacional de Energía
Atómica (CNEA), Universidad Nacional de Cuyo and Consejo Nacional de Investigaciones Científicas y
Técnicas (CONICET), Av. Bustillo 9500, 8400 San Carlos de Bariloche, Argentina, e-mail:
barra@cab.cnea.gov.ar
F. Navarrete, Department of Physics, Kansas State University, Manhattan, KS 66506, USA, e-mail:
navarrete@phys.ksu.edu
M. F. Ciappina, Institute of Physics of the ASCR, ELI-Beamlines, Na Slovance 2, 182 21 Prague,
Czech Republic, e-mail: marcelo.ciappina@eli-beams.eu
Michael Schulz, Department of Physics and LAMOR, Missouri University of Science & Technology,
Rolla, MO 65409, USA, e-mail: schulz@mst.edu

https://doi.org/10.1515/9783110580297-004

Unfortunately (or fortunately, depending on how each of us cope with uncertainties and broken paradigms), when we look deeper into this kind of experiments, we find out that they are trickier than expected. As we will discuss in the present chapter, all and every statement in the previous paragraph is false or at least doubtful. In what follows, we will discuss these issues, paying special attention to the assumption of coherence and lack of contextuality. First, we will explore some basic ideas in very simple terms, which will pave the way for the discussion ahead.

2 Some simple experiments and a historical debate (spoiler alert)

2.1 Throwing pebbles into a pond

Imagine a pond in a calm day. Some scattered reeds protrude from the water near the bank. Now, let us throw a pebble into the water, not far from the reeds. A well-defined set of circular waves will begin to expand through the lagoon. When reaching the reeds, they will be diffracted by them and overlap to give rise to some beautiful interference patterns.

Now let us change this experiment a little bit. Instead of a single pebble, let us throw a handful of pebbles into the pond. A disordered set of waves will reach the reeds, without forming any interference structure. The explanation of this disappointing result is relatively simple. Each individual pebble gives rise to its own and unique interference scheme. However, the random distribution of waves results in an equally random distribution of different interference schemes, and the total effect is completely blurred.

The differences between these two scenarios rely on a concept called "coherence." The first experiment is a typical example of a perfectly coherent system. The second one represents an incoherent superposition of waves. Let us leave this idea maturing for a while, to resume to it later in a more precise way.

Finally, let us walk along the shore and again throw a bunch of pebbles but now far away from the reeds. We will witness an amazing and, perhaps, unexpected result. Before reaching the reeds, the different waves will join in a single circular set of waves, similar to the one created in the first experiment; and the interference patterns will reappear.

Now the situation has become somewhat tricky. We are witnessing how, over time, coherence seems to arise spontaneously from an initially incoherent system. This effect had been known for over two centuries. But a clear understanding and explanation only came to light in the 1930s, when two Dutch physicists, Pieter H. van Cittert [1] and Frits Zernike [2], independently demonstrated a now famous theorem.

2.2 A strip of card illuminated by a sunbeam

Interference effects similar to those observed with water waves can also occur with light. For instance, we all know the famous two-slit experiment described by Thomas Young in his lectures on Natural Philosophy of 1807 [3]. But perhaps we are not so familiar with a previous version, which was presented by Young in the third of his Bakerian lectures in November 1803 [4]. A beam of sunlight enters into a dark room through a pinhole in a window shutter. A narrow strip of card intersects this sunbeam producing an interference pattern on the opposite wall.

At first sight, this experiment looks similar to throwing a pebble into a pond near, in this case, a single reed. A light wave travels from the Sun (i. e., the pebble) to the strip of card (i. e., the reed), producing an interference pattern. However, after giving some thought to this subject, it is clear that we are dealing with the third version of the pond and pebbles' experiment. The sun produces an incoherent superposition of light waves originating in zillions of atomic-size sources. Along its 8 minutes journey toward the Earth, the sunlight is gaining coherence, as it occurs with the waves generated by the bunch of pebbles in the pond. And so, when the sunbeam is intersected by Young's strip of card, it is coherent enough to produce an interference pattern.

But, why do we say that the sunbeam is coherent enough? Is it not purely coherent? Well, it is not. In fact, the strip of card has to be very narrow for the experiment to be successful. This is why the term "coherence length" has been coined. If the strip of card is much larger than the coherence length of sunlight on Earth, the interference pattern will not appear.

Thus, we can now provide a more precise description of the idea that an incoherent superposition of waves gains coherence, by saying that its "coherence length" increases in time. We still lack a physical definition of this length and an explanation of this effect, but at least we are gradually achieving a better understanding.

In what follows, we will be dealing with atomic scattering experiments. They share some similarities with the examples that we have been describing so far. A beam of projectiles (i. e., the waves produced in the pond by the handful of pebbles or the photons in the sunbeam) collides with a target (namely, the reeds in the first case or the strip of card in the second case). But, before moving on, let me ask two extremely simple but transcendent questions. First, is the interference a single-particle or a multi-particle effect? This issue was unambiguously addressed by Dirac in his famous book on Quantum Mechanics [5]. On page 9, he wrote that:

> "Each photon then interferes only with itself. Interference between two different photons never occur."

Rewrite this phrase with massive particles instead of photons, and you have the answer to the previous question. Each pebble thrown into the pond, or each photon in the sunbeam, produces its own interference pattern. But the overlap of different interference patterns from different origins, might blur the final result.

The second silly question reads something like this: Why Young did his experiment inside a dark room with a narrow beam of light, instead of in open air? The answer seems to be quite simple, isn't it? The intensity of the interference pattern is so weak that it would be rendered invisible by daylight. But, is this answer really so obvious? As we will see in the last example, more than 90 years ago this seemingly simple idea baffled the most prominent physicist of their times.

2.3 The 1927 Solvay conference

In the morning of Monday, October 24, 1927, Louis De Broglie delivered the inaugural talk at the fifth Solvay Conference in Brussels, where he presented his Pilot-Wave interpretation of Quantum Mechanics. After his intervention, an interesting discussion broke out. We might say with an obvious double meaning, that the waves set in motion during that discussion have not settled down.

We are not interested in the details of that discussion, but solely on a very specific and particular issue, that was at the root of a famous—and still widely misunderstood—objection raised by Pauli. Actually, most of the discussion concerned the treatment of a scattering process. De Broglie, seconded by Brillouin, had already correctly answered a query raised by Max Born. But at some time in the discussion, Pauli resorted to the common understanding that the incoming projectile should be described by a plane wave, in order to build up an objection against de Broglie's interpretation. The interested reader is referred the book by Bacciagaluppi and Valentini [6] for a description of Pauli's subtle arguments. Here, we are only interested in the initial assumption of an incident plane wave about which, in 1952, David Bohm gave the following opinion [7]:

> "For as is well known, the use of an incident plane wave of infinite extent is an excessive abstraction, not realizable in practice. Actually, both the incident and outgoing parts of ψ-field will always take the form of bounded packets [...] Thus, Pauli's objection is seen to be based on the use of the excessively abstract model of an infinite plane wave."

Contrary to a widespread belief, de Broglie's reply to Pauli did put into evidence that the use of a plane wave made Pauli's objection collapse right from the onset. As he correctly said [6]:

> "The difficulty pointed out by Mr Pauli has an analogue in classical optics. One can speak of the beam diffracted by a grating in an given direction only if the grating and the incident wave are laterally limited, because otherwise all the diffracted beams will overlap and be bathed in the incident wave. [...] One must also assume the wave ψ to be limited laterally in configuration space."

Immediately after this intervention by De Broglie, Lorentz said that *"The question is to know what a particle should do when it is immersed in two waves at the same time."* This phrase clearly shows how confused most of the participants at the conference were, regarding the discussion between De Broglie and Pauli. As an answer to Lorentz's remark, De Broglie repeated like a *mantra* that *"the whole question is to know if one has the right to assume the wave ψ to be limited laterally in configuration space."* [6]

There are at least three group photos which were taken on Tuesday, the 25th, by the Belgian photographer Benjamin Couprie. In one of them, reproduced in Figure 4.1, the hard gesture of Pauli looking straight at De Broglie might be representative of the acrimony left by the discussion of the previous day.

Figure 4.1: Group photo of the Solvay Conference of 1927 by the Belgian photographer Benjamin Couprie. Starting from the right, Pauli is the fourth standing and De Broglie is the third in the second row.

3 The excessively abstract model of an infinite plane wave

And so, almost 90 years later, here we are, still discussing—as David Bohm described it—*the excessively abstract model of an infinite plane wave*. This description of a scattering process can be found in almost "every" textbook of quantum mechanics, as for instance in the standard reference text by Leonard Schiff [8]. He points out that the collimated projectile's beam is not an infinite plane wave. This condition is necessary in order to distinguish the feeble scattered part of the wave (whose amplitude decays as the inverse of the distance to the force center) from the background of the unscattered part. In other words, this is why Young did not perform his experiment in broad daylight. Thus, at the point of observation P in Figure 4.2, only the scattered outgoing part of the wave is present. But, and this is very important, Schiff points out that this scattered outgoing part of the wave is essentially the same that can be calculated assuming an incoming infinite plane wave. So, if this argument is correct, while the experiment requires a collimation, the calculation can be done assuming infinite plane waves. This idea, of course, was very well known and cannot be attributed to Schiff himself. As a reviewer of the first edition of Schiff's book wrote [9]:

> "About a half dozen theoretical physicists with whom the reviewer has spoken have commented on the fact that the text is closely similar to the lecture notes they have employed for a number of years in teaching quantum mechanics to graduate students."

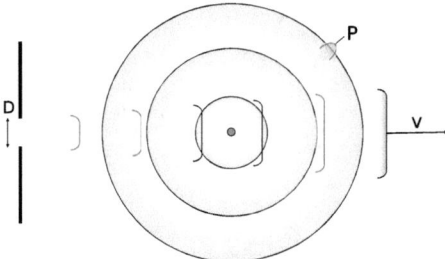

Figure 4.2: Standard model of an elastic collision. A beam of projectiles of velocity v is collimated by a diaphragm of characteristic size D so that at the detection point P only the outgoing scattered wave is present.

In almost every calculation of a scattering cross-section this fundamental and basic ansatz is never questioned, even though, up to the time when Schiff wrote his book, the validity of this assumption had never been proven or at least thoroughly discussed. Apparently, the first analysis was performed by Eyvind Wichmann in 1965 [10]. However, no trace of his demonstration can be found in the textbook on Quantum Mechanics written by Wichmann himself 2 years later [11]. He simply returned to the same old arguments. Years later, Wichmann's analysis was picked up by John Robert Taylor in his well-known textbook on scattering theory [12].

4 Taylor and Wichmann's model in a nutshell

Let us critically review the main characteristics of Taylor and Wichmann's approach. We start by considering a single scattering event (i. e., a single pebble thrown into the pond) as shown in Figure 4.3. We might be studying an elastic or a multichannel process (ionization, charge exchange, etc.), but only its two-particle incoming section would be of interest for us at this point. We can study this problem in simple terms as a one-body problem, where a single particle of "reduced" mass $m = m_P m_T/(m_P + m_T)$ interacts with a force center (e. g., see Chapter 4 in [12]). We characterized this system by a wave packet ψ, which starts at a given macroscopic distance L from the force center and moves toward it with initial impulse **p**. If there were no interaction between the projectile and the target, the orbit would be free and the projectile's position distribution $|\phi|^2$ would look similar to what is shown in Figure 4.3 (gray-shaded area). Here, we are following Taylor in adopting *the classical terminology and refer to $\psi(t)$ as an "orbit."* [12].

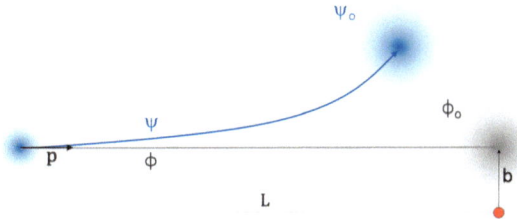

Figure 4.3: The scattering wave packet ψ starts at a given macroscopic distance L from a force center, and moves toward it with initial impulse **p**. We associate to it a free orbit ϕ, and characterize both the scattering and free wave packets by their states ψ_o and ϕ_o at the perihelion.

We characterize this free orbit by its wave packet ϕ_o at its perihelion, when the distance to the force center reaches its minimum. This characteristic distance b is the so-called "impact parameter." In presence of the projectile-target interaction, the reduced particle evolves along a given "orbit" ψ, which we also characterize by its wave function ψ_o at the perihelion. And so we arrive to one of the key assumptions of the standard scattering model, the so-called asymptotic condition, which states that for potentials which fall faster that $1/r$ at large distances, namely $rV(r) \to 0$ for $r \to \infty$, the previous description of a scattering event makes sense, i. e., that at very large distances L from the force center, *any ϕ is in fact the asymptote of some actual orbit ψ* [12]. Furthermore, it is clear that both the scattering and free states at the perihelion can be linearly related by an operator Ω_L, namely $\psi_o = \Omega_L \phi_o$. Finally, since the distance L traveled by the projectiles before reaching the target is macroscopic and, therefore, much larger than any other length associated to the scattering event, we can replace Ω_L by its limit [13] $\Omega_+ = \lim_{L \to \infty} \Omega_L$, as introduced in 1945 by the Danish chemist and physicist, Christian Møller [14]. In this way, we have obtained a direct time-independent relation between the free and the scattering states, allowing for a stationary description of the scattering process.

The second key assumption of the theory is that the collimator size D is also much larger than any characteristic atomic length and, therefore, as long as we keep in mind that the actual detection of the collision products is done in the shadow zone casted by the collimator, its size D can be assumed to be infinite in the calculation of each single scattering event.

Within this two basic assumptions, namely $L \rightarrow \infty$ and $D \rightarrow \infty$, the standard stationary approach of the scattering theory is recovered. As Wichmann wrote in 1965, *the final results obtained are old and well established*. However, Taylor warns us that [12]:

> "Such methods [i. e., using plane waves] can never be regarded as physically satisfactory; but as an introductory route to the correct answer they obviously accomplish an important purpose."

So, let us consider Taylor's warning and analyse the problem somewhat deeper, beyond its "introductory" purpose.

5 Can the distance between the collimator and the target be assumed to be infinite?

It is difficult to imagine how the simple arguments summarily described in the previous section should be wrong. However, a disrupting idea was put forward by Egodapitiya et al. in 2011 [15], by suggesting that *measured cross-sections can sensitively depend on the coherence properties of the projectiles*. This article was soon followed by a series of interesting and challenging experiments (see, e. g., [16] and references therein), which pointed out to the possibility that, in some previously unforseen way, one or both independent limits $L \rightarrow \infty$ and $D \rightarrow \infty$ were not valid. In these experiments, the length L between the collimator and the target, and/or the aperture D of the collimator itself, are modified.

In order to explore this possibility, let us critically review the demonstrations by Wichmann [10] and Taylor [12]. First, let us point out that in the absence of a confining potential, only plane waves don't change in time. All other square-integrable wave packets do. Actually, if we choose the time $t = 0$ as the one at which the full width Δx of a free wave packet reaches its minimum value d [17, 18], its width would spread for both negative and positive times, $\Delta x = \sqrt{d^2 + (2\Delta pt/m)^2}$, with an asymptotic linear dependence $\Delta x \approx 2\Delta p|t|/m$. This simple result clearly poses a difficulty to the previous analysis of a scattering event. In Section 2c of his textbook, Taylor assumed that the zero covariance condition for the free wave packet ϕ occurs at the perihelion [12]. But this assumption leads to an unrealistic situation where ϕ is shrinking while approaching the force center. It seems to make more sense to assume that the contrary occur;

i. e., ϕ spreads until reaching a full width at the perihelion

$$\ell_o = \sqrt{d^2 + \left(\frac{L\lambda}{\pi d}\right)^2},$$ (4.1)

with $\lambda = h/p$ the particle's wavelength. Here, we have assumed that at a distance L from the force center $\Delta x \Delta p$ has its minimum value as allowed by Heisenberg's principle. A similar spreading would also occur for the scattering wave packet ψ.

In conclusion, it seems at least problematic to characterize the scattering and free orbits by their wave packets at the perihelion, since they would unavoidably be dependent on the distance L. However, when we put some numbers, the limit $L \to \infty$ can be recovered under a new light. For instance, an electron wave-packet with $d = 1$ atomic unit would have increased its size up to $\ell_o = 50$ centimetres after traveling a distance of $L = 1$ meter with a velocity of 4 atomic units. This means that the projectile will illuminate the target coherently, and whether we take the limit $L \to \infty$ or not, seems to be of no real concern. But, is it really so? In fact, as we will see in the following section, this assumption is put to test by a missing piece in the description of a scattering experiment.

6 A scattering experiment is not composed of identical events

The previous description of the incoming part of a scattering process was mostly concerned with the analysis of a single wave function ψ, i. e., a single pebble thrown into the pond. In the end, the *old and well-established* [10] infinite plane wave model strengthens the idea that the calculation of a single wave function in the continuum is all that is required to fully achieve a complete description of any collision. However, we have to keep in mind that, as in the case of a handful of pebbles thrown into a pond, this is only one of many events that are collected in an actual experiments. All these events are similar, but not identical. For instance, they might differ in the shape and/or phase of the initial wave function or in the orientation θ of the impulse **p**, just to mention two possible options. But the most conspicuous difference between them is given by their impact parameters b, which are limited by a collimator within certain macroscopic boundaries that we are characterizing by the macroscopic length D (i. e., $b \leq D$), namely the famous *lateral limitation* of the projectile's beam discussed by De Broglie and Pauli in 1927.

Thus, the actual scattering process is composed of many individual events which differ in their orientations θ and impact parameters b, as shown in Figure 4.4. It is very important to clearly understand the differences between these two classical parameters θ and b. The impact parameter b defines the "relative" distance (perpendicular

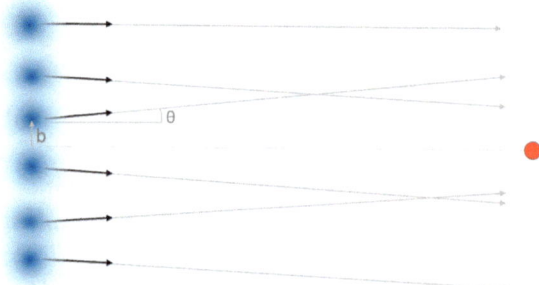

Figure 4.4: The scattering process is composed of many individual events which differ in their orientations θ and impact parameters b. The impact parameters are limited by a collimator within certain boundaries that we are characterizing by a macroscopic length D. Similarly, θ is limited by a maximum angle θ_0 (which we have exaggerated in the figure for a better visualization).

to the experimental set-up axis) between a pair of colliding projectile and target particles. The use of the word "relative" is of the utmost importance. For instance, since the gaseous target is usually spread over a region that is larger than or, at least, comparable with the transversal area of the incoming beam, a projectile at the edge of the collimator can still come into a close encounter (i. e., characterized by a very small impact parameter (b) with a target atom at the edge of the collision region).

The angle θ, on the other hand, is related to the beam divergence. The impulse **p** of each projectile's wave packet might have a "lateral" component $p_\perp$ (again, perpendicular to the experimental set-up axis) so that $\theta \approx p_\perp/p$. Here, $p_\perp$ defines a deviation of the group impulse **p** of the projectile's wave packet and should not be confused with its momentum width $\Delta p_\perp$. While the latter is a quantum quantity, the former has a purely classical origin.

In very simple terms, we can think that while θ_0 is limited by the focusing setup along the equipment, D is solely related to the width of the last collimator; and none of them depends on the distance L to the target. Both θ_0 and D only refer to the initial stage of the collision. These arguments should be kept in mind when, in the following section, we will find it more convenient to employ the dimensionless quantity D/L instead of D and L separately. This quantity can be interpreted as the "angle" subtended by the collimator at the target, and clearly depends on the distance L between them. But let us remember that θ_0, on the other hand, does not depend on L and only describes the divergence angle of the beam.

Both quantities θ and b require a classical statistical averaging. In other words, due to this incomplete knowledge of the initial state, we have to deal with both a quantum and a classical statistic simultaneously. As we previously said, this is similar to what happens when a handful of pebbles is thrown into a pond, where the individual waves evolve in a similar way, but starting at different initial positions. The best way of describing such a system is by employing a density matrix formulation [19]. It provides a uniform treatment of any quantum systems by expressing the maximum information available in a compact way, and allows to perform calculations which would be very cumbersome otherwise.

In the following section, we will avoid the mathematical and physical subtleties of the density matrix formulation, and only show the main results that are obtained by means of this technique.

7 The van Cittert–Zernike theorem revisited

Along the line summarily sketched in the previous section, it is possible to develop a quantum-mechanical analog to what is known in optics as the van Cittert–Zernike theorem [1, 2]. Within this framework, it is demonstrated that while a projectiles beam is moving toward the target, it becomes increasingly coherent, in the sense that it is identical to a pure state within a transversal coherence length given by this simple equation [20, 21]

$$\ell = \sqrt{\frac{d^2}{1 + (\pi \theta_o d/\lambda)^2} + \left(\frac{\lambda L}{\pi \sqrt{d^2 + D^2}}\right)^2}. \tag{4.2}$$

We see that the coherence length consists of the square root of the sum of the squares of the contributions of the angular and impact parameter dispersions, θ_o and D, respectively.

7.1 Impact parameter effects

Let us first assume that $\theta_o = 0$ in order to analyze one of these terms related to the dispersion of the impact parameter along a transversal width of length D, namely $\ell = \sqrt{d^2 + (\lambda L/\pi \sqrt{d^2 + D^2})^2}$. Note that for $D = 0$, i. e., if there were no dispersion on the impact parameter, we recover the coherence length ℓ_o corresponding to a pure state. However, it would be more realistic to assume that the collimator is macroscopically larger than the initial wave packet (i. e., $D \gg d$) so as to approximate the coherence length as follows:

$$\ell_D \approx \sqrt{d^2 + \left(\frac{\lambda L}{\pi D}\right)^2}. \tag{4.3}$$

We see that in this equation, the macroscopic lengths L and D are intertwined in such a way that they cannot be compared solely with atomic size quantities, but only among themselves. Therefore, the limits $L \to \infty$ and $D \to \infty$ cannot be taken separately, as Wichmann and Taylor did. This is the very essence of the van Cittert–Zernike theorem. The remarkable effect here is that the waves emitted from an incoherent source cooperate in such a way as to produce some degree of coherence at large distances. In simpler terms, we are seeing coherence emerging out of an incoherent system.

Finally, in the Fraunhofer limit $D/L \ll \lambda/\pi d$, we obtain $\ell_D \approx \lambda L/\pi D$. This result is similar to the one usually employed in optical studies and, more recently, for analyzing atomic collision processes [16]. When compared with the corresponding Fraunhofer limit for a pure state, $\ell_o \approx \lambda L/\pi d$, we see that the coherence length of a beam is smaller than that of a single projectile by a factor of the order of d/D. In other words, while ℓ_o might reach macroscopic dimensions, this is not necessarily valid for ℓ_D. This is similar to what happens with sunlight, where a single photon would exhibit spatial coherence over the size of the Earth itself, while the coherence length of a sunbeam would be of the order of some tens of micrometers.

In this way, we have found that the coherence length ℓ is the only distance, not L or D, that has to be compared with the characteristic atomic dimensions a of the scattering problem in order to decide whether it can be treated coherently (i. e., with the standard scattering theory) whenever $\ell > a$, or incoherently if $\ell < a$. Actually, by adjusting the ratio D/L, it is possible to switch from a coherent to an incoherent scattering process, an effect which has been extensively investigated both experimentally and theoretically in recent years [16].

7.2 Angular dispersion effects

Now let us concentrate on how the directionality of the projectiles beam might affect the coherence length, as represented by the term, $\ell = d/\sqrt{1 + (\pi\theta_o d/\lambda)^2}$. Let us consider that $\theta_o \gg \lambda/\pi d$, so that $\ell \approx \lambda/\pi\theta_o$. Note the similarity of this expression with $\ell_D \approx \lambda L/\pi D$, where the role of the angle θ_o is played by the angular opening D/L of the collimator as seen from the target position. Actually, if both limits $D \ll d$ and $\theta_o \gg \lambda/\pi d$ apply, then equation (4.2) can be approximated by

$$\ell \approx \frac{\lambda}{\pi}\sqrt{\frac{1}{\theta_o^2} + \frac{1}{(D/L)^2}}. \qquad (4.4)$$

The similarity between the effects produced by the projectile's angular dispersion θ_o and the angular aperture D/L of the collimator has been recently exploited to simulate the modification of the quotient D/L by that of the angle θ_o [22]. However, a word of caution is necessary at this point, since the similarity of θ_o and D/L is only apparent. For instance, coming back to equation (4.2), we see that if θ_o is small enough, then we recover equation (4.1), and varying D/L would make it possible to switch between a total and a partially coherent condition. But the opposite is not true. If D/L is small as to produce a coherent beam, there is no way of changing this condition by modifying θ_0. In other words, the angle D/L subtended by the collimator at the target should have to be much larger than $\lambda/\pi a$, with a a characteristic distance of the target, for the coherence of the beam to be controlled by the angle θ_o alone.

8 Incorporating coherence and contextuality effects in the calculation of differential cross-sections

Through a quantum mechanical equivalent of van Cittert–Zernike theorem, we have shown that a collision experiment might occur coherently or not, depending on whether the coherence length ℓ of the projectiles beam is larger or smaller than the characteristic size a of the atomic system involved, respectively. Furthermore, we have demonstrated that this coherence length is affected by the experimental context, as for instance by the angular dispersion θ_o of the projectiles beam, or the ratio between the size D of the collimator and its distance L to the target.

However, at this point, we still face the problem of how to incorporate these effects in the actual calculation of a cross-section. For instance, for a scattering process leading to an N-body final state, we can calculate the corresponding fully differential cross-section, $d\sigma/d\mathbf{k}_1 \cdots d\mathbf{k}_{N-1}d\mathbf{q}$, by some standard time-independent method [23]. Here, $\mathbf{k}_1, \mathbf{k}_2, \ldots$ are the momenta of each particle in the final state, while $\mathbf{q}$ is the momentum transferred by the projectile.

By applying the ideas of a recent article by Karlovets et al. [24], it has been proposed that the previously discussed coherence effects can be incorporated in the calculation of the actual cross-section by means of a Kernel W in momentum space, as follows [25]:

$$\frac{d\tilde{\sigma}}{d\mathbf{k}_1 \cdots d\mathbf{k}_{N-1}d\mathbf{q}} = \int \frac{d\sigma}{d\mathbf{k}_1 \cdots d\mathbf{k}_{N-1}d\mathbf{q}'} W(\mathbf{q} - \mathbf{q}')d\mathbf{q}. \tag{4.5}$$

It is fair to say that this convolution in momentum space had already been proposed in 2006 by Fiol et al. [26] in order to incorporate experimental uncertainties. Some few years ago, Feagin and Hargreaves [27], used a simplified version of this equation to take into account the degree of collimation of the incident beam. And even more recently, Kouzakov [28] employed it to discuss the effects produced by the target wave packet. Our ansatz is that this same expression can be employed to incorporate coherence effects. It has become customary to choose a Gaussian shape for this kernel,

$$W(\mathbf{q}) \propto \exp -\frac{\gamma}{2\hbar^2}(\mathbf{q}^t \cdot \mathcal{L} \cdot \mathbf{q}). \tag{4.6}$$

Here, $\mathcal{L}$ is a 3×3 matrix whose eigenvalues are the coherence lengths ℓ_i along three orthogonal directions [29]. We employ this generalization, in order to include the possibility of different coherence length in different directions, as it does actually occur in recent experiments (see, e. g., [30]). We have also incorporated a γ factor which depends on the geometry of the collimator [29].

Let us note that by replacing the coherence length (equation (4.2) in equation (4.6)), the kernel W can be written as a product of two independent kernels depending on the contributions of the angular and impact parameter dispersions, θ_o and D/L, respectively.

9 Experimental evidence

Let us start this final section with a timely quotation about the relation between theory and experiment [31]:

> "Some profound experimental work is generated entirely by theory. Some great theories spring from pre-theoretical experiment. Some theories languish for lack of mesh with the real world, while some experimental phenomena sit idle for lack of theory. There are also "happy families," in which theory and experiment coming from different directions meet."

For decades, theoretical perturbative calculations for ion-impact ionization collisions had been agreeing quite fairly with experimental results [23]. Both, theories and experiments had been steadily growing in sophistication, until a major breakthrough was achieved with the development of the Cold Target Recoil–Ion Momentum Spectroscopy (COLTRIMS) technique [32, 33]. The new experimental data obtained by this innovative method were not reproduced by state-of-the-art calculations, with discrepancies that seemed to increase for slower and/or highly charged ions. Thus, the "happy family" engaged in a passionate and interesting discussion over the origin of these discrepancies, that took up most of the first decade of the twenty-first century [16]. Finally, in an experimental study of ionization of Hydrogen molecules by the impact of protons at intermediate energies, Egodapitiya et al. [15] suggested that the coherence properties of the projectiles beam might be playing a major role in the description of ion-impact ionization collisions. Since then, many experiments were performed to validate this disruptive argument, with different theories trying to keep up with the new data [16]. Most of the experiments were based on the control of the parameter D/L, while others dealt with the variation of the angular dispersion θ_o [22]. Note that, in view of equation (4.4), both approaches were assumed to be complementary, in spite of the word of caution given at the end of Subsection 7.2.

Seen in perspective, it is evident that, from the theoretical arena, most of the models developed to explain these results were based in some way or another on a convolution on the momentum transfer as introduced in the previous section. The only difference between these different models seems to lie in the interpretation of this convolution. The theoretical description presented in this chapter tries to summarize most of these interpretations in a single, coherent and comprehensive framework, with a fairly good degree of agreement between experimental and theoretical results.

The detailed description of these experimental and theoretical studies is beyond the scope of the present introductory chapter. The interested reader should resort to other articles cited in the bibliography, in particular to the recent review by Michael Schulz [16]. But let us finish this section by presenting two pieces of evidence on the effects produced by the degree of coherence of the projectiles beam, and the corresponding comparison between theory and experiment.

In Figure 4.5, we show the cross-section for the ionization of H_2 by 75 keV protons, double differential in the solid angle and the energy loss ε of the projectile. In the fig-

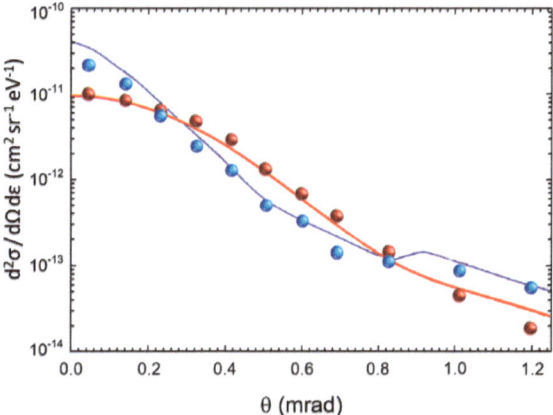

Figure 4.5: Cross-section for the ionization of H_2 by 75 keV protons, double differential in the solid angle and the energy loss of the projectile, as a function of the scattering angle θ. The energy loss ε is fixed at a value of 30 eV. The blue symbols and solid line correspond to the experimental data [15] and theoretical calculations [25] with $D/L \approx 3 \times 10^{-4}$, while the red symbols and curve correspond to $D/L \approx 2.3 \times 10^{-3}$. This latter curve has been divided by a factor of two, in order to highlight the shape agreement between experiment and theory.

ure, ε is fixed at a value of 30 eV [15]. The blue symbols and solid line correspond to the experimental data [15] and theoretical calculations [25] with $D/L \approx 3 \times 10^{-4}$, while the red symbols and curve correspond to $D/L \approx 2.3 \times 10^{-3}$. The difference is that in the first (blue) case, the projectiles beam is coherent enough to illuminate the entire hydrogen molecule. On the other hand, in the second (red) case, ℓ is many times smaller than the internuclear separation of the molecule ($r = 1.4$ au). Remember Young's experiment? We would say that in the first case the coherence length is larger than the width of the strip of card and, therefore, an interference structure is observed. In the second (red) case, however, the opposite occurs, and the interference pattern becomes unobservable. It is fair to point out that in the incoherent (red) case, in spite of a fair agreement in shape, there is still a discrepancy between theory and experiment of a factor of two which could not be accounted for.

The second piece of evidence is shown in Figure 4.6. It is related to the ionization of He by 100 MeV/amu C^{6+} [34] and 1 MeV proton [22] impact. On the upper left panel, which shows the fully differential cross-section (FDCS) in the scattering plane, a calculation (red full line) [29] performed similarly to that in Figure 4.5 (i. e., by taking into account the finite coherence length ℓ of the projectile beam) differ from the completely coherent calculation (blue dashed curve). However, no conclusive evidence can be extracted from the comparison with the experimental data, since both curves agree relatively well with them. The upper right panel show similar results for the case of the 1 MeV proton–Helium collision [22]. No significant differences can be observed between both calculations and with respect to the experimental data. Up to this point,

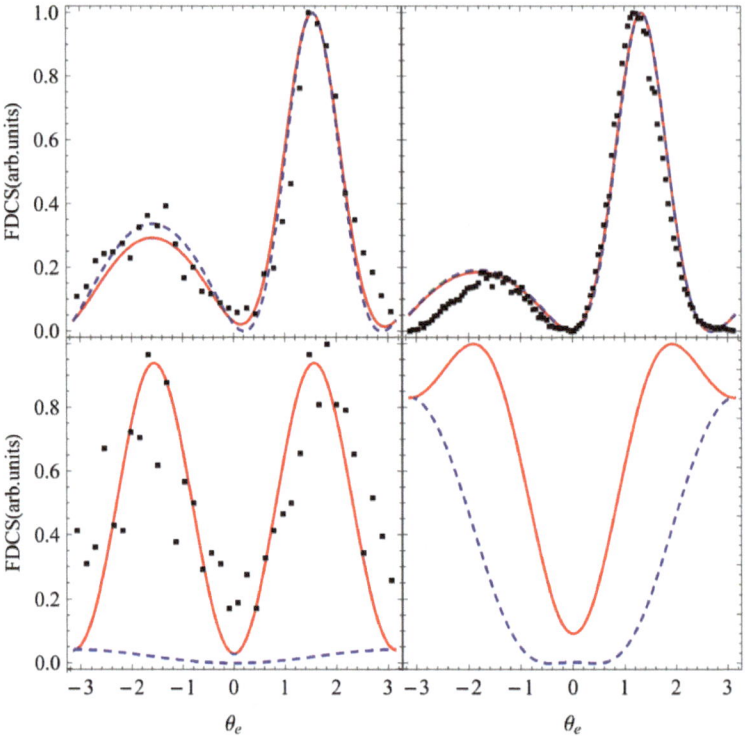

Figure 4.6: Fully differential cross-section (FDCS) for the single ionization of He by 100 MeV/amu C^{6+} ions (left panels) and 1 MeV protons (right panels). In the upper panels, the curves are arbitrarily normalized to the maxima of the experimental data distributions. In the lower panels, the same normalization of the theoretical curves is employed, independently of the experimental data. The FDCS is represented in the scattering (upper panels) and perpendicular (lower panels) planes, as a function of the ejected electron polar angle θ_e, for a fixed ejected electron energy E_e = 6.5 eV. The blue dashed line shows the FDCS calculated in the traditional coherent form. The red (solid) line takes into account the coherence length of the projectiles beam [29]. The black dots correspond to the measurements from [34] in the left panels and [22] in the right panels.

the results are clearly inconclusive. However, when the FDCS for the 100 MeV/amu C^{6+} case is observed in the perpendicular plane, as shown in the lower left panel, the calculation performed with a finite coherence length (red full line) shows a strikingly good agreement with the experimental data. The coherent calculation (blue dashed curve), on the other hand, completely fails to reproduce the experimental results. A similar and important difference between the fully and partially coherent cases is observed for 1 MeV proton–Helium collisions, as shown in the lower right panel. Unfortunately, this plane was not experimentally analyzed [22].

Thus, it seems that the theoretical analysis presented in this chapter is consistent with the experimental results. However, much more work is still to be done before this "happy family" comes to a complete agreement.

10 Conclusions

Since the advent of atomic physics, it has been customarily assumed that scattering experiments were mostly independent of the experimental context. Later on, this ansatz was substantiated by Wichmann [10] and Taylor [12], who demonstrated how to transform what is basically a time-dependent process into a stationary one, where the incoming projectiles can be asymptotically represented by plane waves.

However, when this standard scattering theory is critically reviewed in terms of the density matrix formulation, which by the way is the only correct form of dealing with a problem which includes classical and quantum components, a different picture emerges. The projectiles constitute a conspicuous incoherent beam, but it can build up coherence while approaching the target. If, when the actual collision occurs, the coherence length is larger than the characteristic atomic dimensions, then the standard scattering theory can be employed. Otherwise, the contextuality of the experiment has to be taken into account.

This contextuality might intervene in different ways as, for instance, in the directionality of the incoming beam or the different impact parameters limited by a collimator. These effects can be incorporated in the actual calculation of the cross-section by means of a convolution in the momentum transfer. Both the angular and impact parameter dispersions can be experimentally controlled, making it possible to switch from a coherent to an incoherent scattering process (see [16] and references therein).

It is important to point out that these techniques should not be mistakenly understood as a deliberate worsening of the experimental uncertainties intended to reduce the coherence length. Actually, instead of being related to the detection stage, they represent a controlled preparation of the initial set-up.

These new possibilities, which were overlooked for decades, are opening unexplored experimental and theoretical possibilities for the study of coherence effects in atomic collisions. The future will tell where the far-reaching implications of these new and promising lines of research would lead us, in a scientific exploration which is not alien to the very foundations of quantum physics.

Bibliography

[1]   van Cittert PH. Die Wahrscheinliche Schwingungsverteilung in Einer von Einer Lichtquelle Direkt Oder Mittels Einer Linse Beleuchteten Ebene. Physica. 1934;1:201–210.
[2]   Zernike F. The concept of degree of coherence and its application to optical problems. Physica. 1938;5:785–795.
[3]   Young T. A Course of Lectures on Natural Philosophy and the Mechanical Arts. London: J. Johnson; 1807.
[4]   Young T. Experiments and calculations relative to physical optics. Philos. Trans. R. Soc. London. 1804;94:1–16.
[5]   Dirac PAM. The Principles of Quantum Mechanics. Oxford University Press; 1930.

[6] Bacciagaluppi G, Valentini A. Quantum Theory at the Crossroads. Reconsidering the 1927 Solvay Conference. Cambridge: Cambridge University Press; 2009.

[7] Bohm D. A Suggested Interpretation of the Quantum Theory in Terms of "Hidden" Variables. II. Phys. Rev. 1952;85:180–193.

[8] Schiff L, Quantum Mechanics. 1st edn. New York: McGraw Hill; 1949.

[9] Steitz F. Review: L. I. Schiff, Quantum Mechanics. Bull. Amer. Math. Soc. 1950;56:191–192.

[10] Wichmann EH. Scattering of Wave Packets. Amer. J. Phys. 1965;33:20–31.

[11] Wichmann EH. Quantum Physics: Berkeley Physics Course. Vol 4. New York: McGraw Hill; 1967.

[12] Taylor JR. Scattering theory: the quantum theory on nonrelativistic collisions. New York: Wiley; 1972.

[13] Belinfante FJ, Møller C. On the relation between the time-dependent and stationary treatments of collision processes. Det Kgl. Danske Videnskabernes Selskab Matematisk-Fysiske Meddelelser. 1954;28(6).

[14] Møller C. General Properties of the Characteristic Matrix in the Theory of Elementary Particles I. Det Kgl. Danske Videnskabernes Selskab Matematisk-Fysiske Meddelelser. 1945;23(1).

[15] Egodapitiya KN, Sharma S, Hasan A, Laforge AC, Madison DH, Moshammer R, Schulz M. Manipulating Atomic Fragmentation Processes by Controlling the Projectile Coherence. Phys. Rev. Lett. 2011;106:153202.

[16] Schulz M. The Role of Projectile Coherence in the Few-Body Dynamics of Simple Atomic Systems. Adv. Atom. Mol. Opt. Phys. 2017;66:507–543.

[17] Robinett RW, Doncheski MA, Bassett LC. Simple Examples of Position-Momentum correlated Gaussian Free-Particle Wave Packets in One Dimension with the General Form of the Time-dependent Spread in Position. Found. Phys. Lett. 2005;18:455–575.

[18] More precisely, we assume that at $t = 0$ the generalized covariance, $\mathrm{cov}(x, p, t) = \frac{1}{2}\langle [x - \langle x \rangle][p - \langle p \rangle] + [p - \langle p \rangle][x - \langle x \rangle]\rangle$ is zero.

[19] Blum K, Density Matrix Theory and Applications. Heidelberg: Springer; 1981.

[20] Fabre I, Navarrete F, Sarkadi L, Barrachina RO. Loss of wave-packet coherence in ion-atom collisions. Eur. J. Phys. 2018;39:015401.

[21] This result generalizes the one described in [20] by incorporating a spread of angle θ_o in the orientation of the impulse **p**.

[22] Gassert H, Chuluunbaatar O, Waitz M, Trinter F, Kim H-K, Bauer T, Laucke A, Müller C, Voigtsberger J, Weller M, Rist J, Pitzer M, Zeller S, Jahnke T, Schmidt LP H, Williams JB, Zaytsev SA, Bulychev AA, Kouzakov KA, Schmidt–Böcking H, Dörner R, Popov YV, Schöffler MS. Agreement of Experiment and Theory on the Single Ionization of Helium by Fast Proton Impact. Phys. Rev. Lett. 2016;116:073201.

[23] Stolterfoht N, Dubois N, Rivarola RD. Electron Emission in Heavy Ion-Atom Collisions. Springer Series on Atomic, Optical and Plasma Physics, 20. Berlin: Springer-Verlag; 1997.

[24] Karlovets DV, Kotkin GL, Serbo VG. Scattering of wave packets on atoms in the Born approximation. Phys. Rev. A. 2015;92:052703.

[25] Sarkadi L, Fabre I, Navarrete F, Barrachina RO. Loss of wave-packet coherence in ion-atom collisions. Phys. Rev. A. 2016;93:032702.

[26] Fiol J, Otranto S, Olson RE. Critical comparison between theory and experiment for C^{6+} + He fully differential ionization cross sections. J. Phys. B. 2006;39:L285–L290.

[27] Feagin JM, Hargreaves L. Loss of wave-packet coherence in stationary scattering experiments. Phys. Rev. A. 2013;88:032705.

[28] Kouzakov KA. Theoretical analysis of the projectile and target coherence in COLTRIMS experiments on atomic ionization by fast ions. Eur. Phys. J. D. 2017;71:70613–70617.

[29] Navarrete F, Ciappina MF, Sarkadi L, Barrachina RO. The role of the wave packet coherence on the ionization cross section of He by p^+ and C^{6+} projectiles. Nucl. Instr. Meth. B.

2017;408:165–168.

[30] Lamichhane BR, Arthanayaka T, Remolina J, Hasan A, Ciappina MF, Navarrete F, Barrachina RO, Lomsadze RA, Schulz M. Fully Differential Study of Capture with Vibrational Dissociation in p + H−2 Collisions. Phys. Rev. Lett. 2017;119:083402.

[31] Hacking I, Representing and intervening. Introductory topics in the philosophy of natural science. Cambridge University Press; 1983.

[32] Dörner R, Mergel V, Jagutzski O, Spielberger L, Ullrich J, Moshammer R, Schmidt-Böcking H. Cold target recoil Ion momentum spectroscopy: a 'Momentum Microscope' to view atomic collision dynamics. Phys. Rep. 2000;330:95.

[33] Ullrich J, Moshammer R, Dorn A, Dörner R, Schmidt L, Schmidt-Böcking H. Recoil-ion and electron momentum spectroscopy: reaction-microscopes. Rep. Prog. Phys. 2003;66:1463.

[34] Schulz M, Moshammer R, Fischer D, Kollmus H, Madison DH, Jones S, Ullrich J. Three-dimensional imaging of atomic four-body processes. Nature. 2003;422:48.

A. B. Voitkiv

5 A distorted-wave approach to projectile-electron excitation and loss in high-energy ion-atom collisions

1 Introduction

A large number of different atomic physics processes may occur when an ion (projectile), which carries initially electron(s), collides with a neutral atom (target). In this chapter, we shall deal with the process of projectile-electron loss (or excitation) in collisions with atoms in its very basic version in which only one "active" electron of the projectile participates.

This process will be considered for the high-energy collision range which is defined by the condition that the impact velocity v greatly exceeds the typical orbiting velocities $v_0 \simeq e^2/\hbar$ of outer-shell atomic electrons: $v_0 \ll v < c$, where $c = v_0/\alpha$ is the speed of light in vacuum ($\alpha = e^2/\hbar c \approx 1/137$ is the fine-structure constant). The focus will be on electron loss from (excitation of) highly charged ions and on how this process can be described using distorted-wave models.

A theoretical consideration of ion-atom collisions often exploits a huge difference between masses of electrons and nuclei. Because of this, the nuclei move in the collision practically undeflected and can be regarded as the source of an external field in which the electrons undergo transitions. In *semiclassical approximation*, only the electrons are treated as quantum particles whereas the nuclei are supposed to move along classical trajectories (see, e. g., [1]). One normally chooses a reference frame in which one of the nuclei is at rest and takes its position as the origin. The other nucleus moves is this frame along a straight-line trajectory $\mathbf{R}(t) = \mathbf{b} + \mathbf{v}t$, where $\mathbf{b}$ is the impact parameter, $\mathbf{v}$ the collision velocity and t is the time.

Within the semiclassical approximation, one obtains the semiclassical transition amplitude $a_{fi}(\mathbf{b})$. For collisions with very small scattering angles of the incident nucleus and negligible recoil velocities of the nucleus, which is initially at rest, (both these conditions are very well fulfilled in high-energy collisions), the transition amplitude $S_{fi}(\mathbf{q}_\perp)$, obtained using a fully quantum treatment, is related to $a_{fi}(\mathbf{b})$ by the Fourier transformation

$$S_{fi}(\mathbf{q}_\perp) = \frac{1}{2\pi} \int d^2\mathbf{b}\, a_{fi}(\mathbf{b}) \exp(i\mathbf{q}_\perp \cdot \mathbf{b}), \tag{5.1}$$

where $\mathbf{q}_\perp$ is the transverse part ($\mathbf{q}_\perp \cdot \mathbf{v} = 0$) of the total momentum transfer $\mathbf{q}$ in the collision.

A. B. Voitkiv, Institut für Theoretische Physik I, Heinrich Heine University of Düsseldorf, Universitätsstrasse 1, 40225 Düsseldorf, Germany, e-mail: alexander.voitkiv@tp1.uni-duesseldorf.de

https://doi.org/10.1515/9783110580297-005

The process of projectile-electron excitation or loss in collisions with atoms involves at least four particles and its detailed dynamic description is far from being simple. However, important information about momentum transfers in the process can already be extracted by considering just its kinematics that enables one to obtain the minimum momentum transfer, which represents the component of the total momentum transfer parallel to the collision velocity. In particular, the minimum momentum transfer q^I_{min} to the ion (given in the rest of the ion) and the minimum momentum transfer q^A_{min} to the atom (given in the rest frame of the atom) read (see, e. g., [2])

$$q^I_{min} = \frac{\varepsilon_f - \varepsilon_i}{v} + \frac{\epsilon_f - \epsilon_i}{\gamma v}$$

$$q^A_{min} = \frac{\varepsilon_f - \varepsilon_i}{v\gamma} + \frac{\epsilon_f - \epsilon_i}{v}. \tag{5.2}$$

Here, ε_i (ε_f) is the energy of the initial (final) state of the electron of the ion given in its rest frame, ϵ_i (ϵ_f) is the energy of the initial (final) state of the atom given in its rest frame, and $\gamma = 1/\sqrt{1 - v^2/c^2}$ is the collisional Lorentz factor.

The simplest consideration of projectile-electron loss (or excitation) is based on the first order of perturbation theory in the interaction between the ion and atom—the (first) Born approximation. It is valid if the colliding particles represent comparatively weak perturbations for each other that is (on overall) the case provided

$$\frac{Z}{v} \ll 1 \quad \left(\frac{Ze^2}{\hbar v} \ll 1\right), \tag{5.3}$$

where $Z = \max\{Z_I, Z_A\}$, and Z_I and Z_A are the atomic numbers of the ion and atom, respectively.

If the condition (5.3) is violated the Born approximation becomes invalid and better theoretical methods have to be used. In such a case, distorted-wave models often represent one of the best choices.

The basic idea of a distorted-wave approach to collisions is rather simple. In a standard perturbation theory, the Hamiltonian of the system is split into a free part, describing noninteracting atom and ion, and the interaction between them. In a distorted-wave approach the Hamiltonian is also split into two parts. However, one now tries to include as much of the ion-atom interaction as possible already into the definition of the initial and final states treating the remaining part of the Hamiltonian as a perturbation. Since this residual perturbation usually turns out to be substantially weaker than the ion-atom interaction, a distorted-wave series has a better convergence and it often suffices to consider just its first term.

In the next sections, we discuss distorted-wave models which are applied to describe electron transitions in a highly charged ion caused by collisions with many-electron (heavy) atoms (Sections 2 and 3) and with very light atoms (Sections 4 and 5).

Atomic units are used throughout unless otherwise stated.

2 High-energy collisions with many-electron atoms I

In high-energy collisions with many-electron (heavy) atoms ($Z_A \gg 1$), the main effect of the atomic electrons on electron transitions in the projectile is to screen the field of the atomic nucleus [2]. In collisions of highly charged projectiles with many-electron atoms occurring at not too high impact energies, which can be characterized by the condition

$$\frac{Z_I^2}{\gamma v} \gg Z_A^{1/3}, \tag{5.4}$$

the momentum transfers to the atom in the collision (its rough estimate is given by the left-hand side of (5.4)) is so large that only the impact parameters, which are quite small on the scale of a neutral atom, noticeably contribute to the loss process (see [2] and references therein). In such collisions, provided the ratio Z_A/v is not much smaller than 1, the field of the atomic nucleus becomes too strong to be described within the Born approximation but the role of the atomic electrons is negligible. Therefore, the problem of projectile-electron loss (or excitation) in collisions with a neutral atom can be reduced to the problem of electron transitions in a three-body system (the electron of the ion and the nuclei of the ion and atom) and then treated by using three-body coulomb distorted-wave models (see [3], [2], and references therein). For simplicity in this section, we briefly discuss some of these models only for the nonrelativistic domain of the collision parameters.

In a reference frame, where the nucleus of the ion is at rest and taken as the origin, the Schrödinger equation in the semiclassical approximation reads

$$i\frac{\partial}{\partial t}\Psi = \hat{H}\Psi \tag{5.5}$$

where the semiclassical Hamiltonian $\hat{H}$, which explicitly depends on time, is given by

$$\hat{H} = \hat{H}_0 - \frac{Z_A}{|\mathbf{r} - \mathbf{R}_A|} + \frac{Z_I Z_A}{R_A}. \tag{5.6}$$

Here, $\hat{H}_0$ is the Hamiltonian for a free (noninteracting) ion,

$$\hat{H}_0 = \frac{\hat{\mathbf{p}}^2}{2} - \frac{Z_I}{r}, \tag{5.7}$$

where $\mathbf{r}$ and $\hat{\mathbf{p}}$ are the electron coordinates and momentum operator, respectively. The last two terms in equation (5.6) describe the interaction of the nucleus of the atom moving along the trajectory $\mathbf{R}_A(t) = \mathbf{b} + \mathbf{v}t$ with the electron and the nucleus of the ion, respectively.

The post and prior forms of the semiclassical transition amplitude $a_{fi}(\mathbf{b})$ are given by

$$a_{fi}^{\text{post}}(\mathbf{b}) = -i \int\limits_{-\infty}^{+\infty} dt \left\langle \left(\hat{H} - i\frac{\partial}{\partial t} \right) \phi_f(t) \middle| \Psi^{(+)}(t) \right\rangle \tag{5.8}$$

and

$$a_{fi}^{\text{prior}}(\mathbf{b}) = -i \int\limits_{-\infty}^{+\infty} dt \left\langle \Psi^{(-)}(t) \middle| \left(\hat{H} - i\frac{\partial}{\partial t} \right) \chi_i(t) \right\rangle, \tag{5.9}$$

respectively (see, e. g., [4]). Here $\Psi^{(-)}$ ($\Psi^{(+)}$) is the solution of the Schrödinger equation (5.5) with the full semiclassical Hamiltonian (5.6) satisfying the "in" ("out") boundary condition, $\chi_i(t)$ ($\phi_f(t)$) is the initial (final) state of the electron.

In distorted-wave models, which are intended to describe electron transitions occurring basically within the same colliding center, χ_i and $\Psi^{(-)}$ (or $\Psi^{(+)}$ and ϕ_f) are approximated by

$$\chi_i = L_i \zeta_0 \exp(-i\varepsilon_0 t)$$
$$\Psi^{(-)} = L_f \zeta_\mathbf{p} \exp(-i\varepsilon_\mathbf{p} t). \tag{5.10}$$

Here, ζ_0 and $\zeta_\mathbf{p}$ are eigenstates of the Hamiltonian (5.7), where ζ_0 is the initial (ground) state of the electron in the free ion and $\zeta_\mathbf{p}$ describes (in case of electron loss) the emitted electron moving in the field of the nucleus of the ion with an asymptotic momentum $\mathbf{p}$. Further, L_i and L_f are so called distortion factors for the initial and final states, respectively, of the electron of the ion which describe the distortion of these states caused by the interaction with the atomic nucleus.

The first Born approximation can be recovered from (5.10) by setting there $L_i = L_f = 1$. Accordingly the interaction, which couples the initial and final states in the first Born amplitude, is given by $\hat{H} - \hat{H}_0 = -\frac{Z_A}{|\mathbf{r}-\mathbf{R}_A|} + \frac{Z_A Z_I}{|\mathbf{R}_A|}$, and is the "physical" interaction between the ion and the atom. Although the Born approximation is not exact, its post and prior forms coincide.

Unlike the Born approximation, in distorted-wave models the post and prior forms of the transition amplitude in general differ. Besides, since in these models the projectile-target interaction is already partly included into the states (5.10), the residual interaction, which causes the transitions, differs from the "physical" one of the Born approximation.

CDW-EIS. In the so-called continuum-distorted-wave–eikonal-initial-state (CDW-EIS) model, which was proposed in [5] for ionization of atoms by fast charged particles, the distortion factors are chosen according to

$$L_i = L_i^{\text{eik}} = \exp\left(-iZ_A Z_I \int\limits_{-\infty}^{t} dt' 1/R_A(t') \right)$$

$$\times \exp\left(iZ_A \int\limits_{-\infty}^{t} dt' 1/s(t') \right)$$

$$L_f = L^{\text{cdw}} = \exp\left(-iZ_A Z_I \int\limits_{+\infty}^{t} dt' 1/R_A(t') \right)$$

$$\times \exp(\pi Z_A/(2\kappa))\Gamma(1 + iZ_A/\kappa)$$

$$\times_1 F_1(-iZ_A/\kappa; 1; -i\kappa s - i\boldsymbol{\kappa} \cdot \mathbf{s}), \tag{5.11}$$

where $\mathbf{s} = \mathbf{r} - \mathbf{R}_p$, $\boldsymbol{\kappa} = \mathbf{p} - \mathbf{v}$ is the relative velocity of the emitted electron with respect to the atomic nucleus, Γ is the gamma-function and $_1F_1$ is the confluent hypergeometric function (see, e. g., [6]).

SEA. In the symmetric eikonal approximation (SEA) (which was initially suggested for electron capture in [7]), the distortion factors are taken as

$$L_i = L_i^{\text{eik}} = \exp\left(-iZ_A Z_I \int\limits_{-\infty}^{t} dt' 1/R_A(t') \right)$$

$$\times \exp\left(iZ_A \int\limits_{-\infty}^{t} dt' 1/s(t') \right)$$

$$L_f = L_f^{\text{eik}} = \exp\left(-iZ_A Z_I \int\limits_{+\infty}^{t} dt' 1/R_A(t') \right)$$

$$\times \exp\left(iZ_A \int\limits_{+\infty}^{t} dt' 1/s(t') \right). \tag{5.12}$$

Comparing equations (5.12) with equations (5.11), we see that in both models the distortion factor for the initial state is the same. However, the full Coulomb distortion factor in the final state of the CDW-EIS is replaced in the SEA by its asymptotic (eikonal) form.

Since the semiclassical transition amplitude depends on the impact parameter, which is not an observable quantity, it may not always be used. For instance, the basic dynamics of the collision (the fully differential cross section) cannot be treated with such an amplitude. However, then one can make use of the relation (5.1) and obtain the quantum transition amplitude. The latter is a function of the total momentum transfer in the collision $\mathbf{q}$, which within three-body models is given by

$$\mathbf{q} = \left(\mathbf{q}_\perp, \frac{\varepsilon_p - \varepsilon_0}{v} \right), \tag{5.13}$$

and enables one to calculate any kind of the cross-section.

Since the semiclassical and quantum transition amplitudes are related by the Fourier transformation, the following identity holds (Parseval's theorem):

$$\int d^2\mathbf{q}_\perp |S_{fi}(\mathbf{q}_\perp)|^2 = \int d^2\mathbf{b} |a_{fi}(\mathbf{b})|^2. \tag{5.14}$$

The CDW-EIS and SEA models have been successfully used for calculating such quantities as the total cross-sections and energy-angular distributions of the emitted electrons. As it follows from equation (5.14), these quantities can be obtained either by integrating the absolute square of the quantum amplitude over the transverse momentum transfer or the absolute square of the semiclassical transition amplitude over the impact parameter.

The factors L_i and L_f in equations (5.11) and (5.12) include also the distortion due to the interaction between the nuclei of the colliding particles. However, this distortion appears there just as a phase factor independent of the electron coordinates and after squaring $a_{fi}(\mathbf{b})$ the result already does not depend on this interaction. Consequently, this interaction does not influence cross sections which are obtained by integrating over the impact parameter (or the transverse momentum transfer) and can simply be omitted when such cross sections are calculated. The physical reason for this is that in high-energy ion-atom collisions the interaction between the nuclei practically does not change their motion in the reaction zone. As a result, any indirect coupling between the electron and the nuclei, which would be caused by the interaction between the nuclei, is practically absent.

Three-body Coulomb distorted-wave models can be generalized to the relativistic domain of ion-atom collisions where neither the collision velocity v nor the typical orbiting velocity v_{orb} of the electron in the initial state of a highly charged ion are supposed to be much smaller than the speed of light. Relativistic distorted-wave models are discussed in [3] and [2]. Figure 5.1 gives an example of the application of such models to the projectile-electron loss process in the relativistic domain.

3 High-energy collisions with many-electron atoms II

With the increase in the impact energy, larger and larger impact parameters start to contribute to the projectile-electron loss and at some point the presence of the atomic electrons – in particular, their screening effect – can no longer be ignored [2]. In collisions with many-electron atoms, the screening effect becomes substantial when

$$\frac{\gamma v}{Z_I^2} \gtrsim \frac{1}{Z_A^{1/3}} \tag{5.15}$$

With a further increase in the impact energy, when

$$\frac{\gamma v}{Z_I^2} \gtrsim 1, \tag{5.16}$$

the screening effect becomes of extreme importance qualitatively changing the dependence of the loss cross section on the collision energy (see [2]).

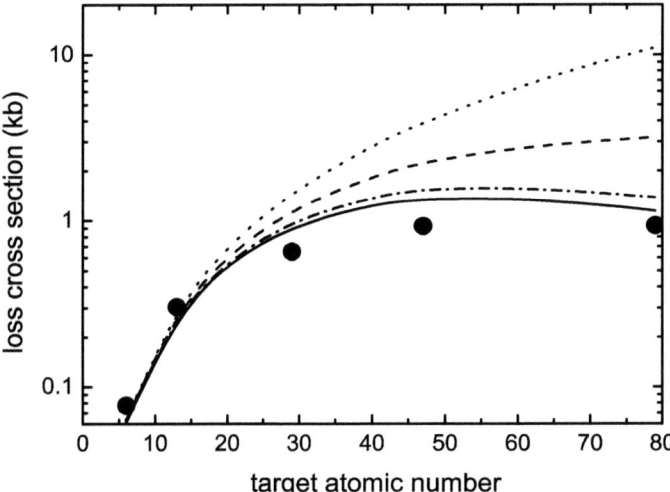

Figure 5.1: The total cross section (per electron) for the single electron loss from 105 MeV/u $U^{90+}(1s^2)$ ions colliding with different targets. Circles show experimental results for the loss in collisions with solid state targets of beryllium, carbon, aluminum, copper, silver, and gold which were measured in [8]. Dot curve show first Born calculations, dash, dash-dot, and solid curves display results of various distorted-wave calculations (see [9]).

Following first-order considerations of projectile-electron transitions, one normally distinguishes two atomic modes in which atomic electrons are "passive" or "active" in the collision process (see, e. g., [10], [11], [2]). In the "passive" mode, atomic electrons, remaining in their initial state, act coherently with the atomic nucleus that leads to a partial or full screening of the field of the latter. In the "active" mode, atomic electrons themselves undergo transitions and, according to first-order treatments, it is the interaction of these electrons with the projectile electron which is fully responsible for projectile-electron transitions in this mode.

When the ion-atom interaction becomes stronger and first-order considerations break down, the classification of atomic modes (and the roles of atomic electrons in them) becomes not so simple. For instance, it is unlikely that in collisions leading to electron loss from a highly charged ion the atomic electrons will remain finally in their initial state (no matter whether they "actively" participate in the interaction with the electron of the highly charged ion or not). However, what is still plausible to assume in such a case [12] is that the shape of the initial electron cloud will not noticeably change during the very short effective collision time and that the coherent action of the atomic nucleus and the "frozen" atomic electron cloud will be mainly responsible for electron transitions in the highly charged ion in collisions with many-electron atoms. Therefore, the process of electron transitions in the ion can again be reduced to a three-body problem in which the atom consisting of its nucleus and electrons is regarded as an external field represented by a superposition of the field of the atomic nucleus and the field of the atomic electrons whose space distribution is "frozen" during the very short collision time.

In the rest frame of the ion, its electron is described by the Dirac equation

$$i\frac{\partial \Psi(\mathbf{r}, t)}{\partial t} = \hat{H}\Psi(\mathbf{r}, t). \tag{5.17}$$

The relativistic Hamiltonian $\hat{H}$ reads

$$\hat{H} = \hat{H}_0 + \hat{W}, \tag{5.18}$$

where

$$\hat{H}_0 = c\boldsymbol{\alpha} \cdot \mathbf{p} + \beta c^2 - \frac{Z_I}{r} \tag{5.19}$$

is the Hamiltonian for the electron in the free (noninteracting) ion and

$$\hat{W} = \boldsymbol{\alpha} \cdot \mathbf{A}(\mathbf{r}, t) - \Phi(\mathbf{r}, t) \tag{5.20}$$

is the interaction between the electron and the field of the atom. Further, $\boldsymbol{\alpha}$ and β are the Dirac's matrices (see, e. g., [13]), $\mathbf{r}$ coordinates of the electron with respect to the ionic nucleus, Φ and $\mathbf{A}$ are, respectively, the scalar and vector potentials of the electromagnetic field generated by the incident atom.

The field of the atom in its rest frame is represented by the scalar potential which, using results of [14] and [15], can be approximated by

$$\Phi' = \frac{Z_A \phi(r')}{r'}, \tag{5.21}$$

where

$$\phi(r') = \sum_{j=1}^{3} A_j \exp(-\kappa_j r') \tag{5.22}$$

with the screening parameters A_j ($\sum_j A_j = 1$) and κ_j tabulated in [14] and [15]. In the rest frame of the ion, in which the atom moves along a classical straight-line trajectory $\mathbf{R}_A = \mathbf{b} + \mathbf{v}t$ ($\mathbf{b} = (b_x, b_y)$, $\mathbf{v} = (0, 0, v)$), the potentials of the field of the atom are obtained from equations (5.21)–(5.22) by a Lorentz transformation that yields [12]

$$\Phi(\mathbf{r}, t) = \frac{\gamma Z_A}{\sqrt{\gamma^2(z - vt)^2 + (\mathbf{r}_\perp - \mathbf{b})^2}} \sum_j A_j \exp\left(-\kappa_j \sqrt{\gamma^2(z - vt)^2 + (\mathbf{r}_\perp - \mathbf{b})^2}\right).$$

$$\mathbf{A}(\mathbf{r}, t) = \left(0, 0, \frac{v}{c}\Phi\right) \tag{5.23}$$

where $\mathbf{r} = (\mathbf{r}_\perp, z)$ with $\mathbf{r}_\perp \cdot \mathbf{v} = 0$.

Within the relativistic symmetric eikonal model, the semiclassical transition amplitude is approximated by

$$a_{fi}(\mathbf{b}) = -i \int\limits_{-\infty}^{+\infty} dt \langle \chi_f(t) | (\hat{H} - i\partial/\partial t) \chi_i(t) \rangle. \tag{5.24}$$

The initial and final states of the electron, which are distorted by the field of the atom, are taken as

$$\chi_i(t) = \psi_0 \exp(-i\varepsilon_0 t) \exp\left(i \int\limits_{-\infty}^{t} dt' \Phi(t') \right)$$

$$\chi_f(t) = \psi_n \exp(-i\varepsilon_n t) \exp\left(i \int\limits_{+\infty}^{t} dt' \Phi(t') \right), \tag{5.25}$$

where ψ_0 and ψ_n are the initial and final undistorted states of the electron in the ion.

The dependence of the scalar potential Φ on the electron coordinates and time is of the form $\Phi = \gamma Z_A f(s_\perp, \gamma | z - vt |)$, where $\mathbf{s}_\perp = \mathbf{r}_\perp - \mathbf{b}$. This can be exploited to transform the amplitude (5.24) into

$$a_{fi}(\mathbf{b}) = i \frac{c}{v} \int\limits_{-\infty}^{+\infty} dt \, \exp(i\omega_{n0} t) \langle \psi_n | \exp\left(i \int\limits_{-\infty}^{+\infty} dt' \Phi(t') \right)$$

$$\times \left(\frac{\Phi(t)}{\gamma^2} \alpha_z - v \left(\nabla_\perp \int\limits_{-\infty}^{t} dt' \Phi(t') \right) \cdot \boldsymbol{\alpha}_\perp \right) | \psi_0 \rangle, \tag{5.26}$$

where $\omega_{n0} = \varepsilon_n - \varepsilon_0$ is the electron transition frequency and $\nabla_\perp$ denotes the two-dimensional (in the (x, y)-plane) gradient operator. Expression (5.26) can be somewhat simplified by noting that

$$\lim_{\lambda \to +0} \int\limits_{-\infty}^{+\infty} dt \, \exp(i\omega_{n0} t) \exp(-\lambda |t|) \int\limits_{-\infty}^{t} dt' \Phi(t')$$

$$= \frac{i}{\omega_{n0}} \int\limits_{-\infty}^{+\infty} dt \, \exp(i\omega_{n0} t) \Phi(t), \quad (\omega_{n0} \neq 0), \tag{5.27}$$

which yields

$$a_{fi}(\mathbf{b}) = i \frac{c}{v} \int\limits_{-\infty}^{+\infty} dt \, \exp(i\omega_{n0} t) \langle \psi_n | \exp\left(i \int\limits_{-\infty}^{+\infty} dt' \Phi(t') \right)$$

$$\times \left(\frac{\Phi(t)}{\gamma^2} \alpha_z - i \frac{v}{\omega_{n0}} (\nabla_\perp \Phi(t)) \cdot \boldsymbol{\alpha}_\perp \right) | \psi_0 \rangle. \tag{5.28}$$

If the interaction between the electron of the ion and the atom is sufficiently weak, one can replace $\exp(i \int_{-\infty}^{+\infty} dt' \Phi(t'))$ in (5.28) by 1. After that we integrate in (5.28) the term proportional to $(\nabla_{\perp}\Phi(t)) \cdot \boldsymbol{\alpha}_{\perp}$ by parts over the (x, y)-plane and use the continuity equation

$$\frac{\partial \rho_{n0}}{\partial t} + \nabla \cdot \mathbf{j}_{n0} = 0 \tag{5.29}$$

for the transition charge and current densities,

$$\rho_{n0} = -\psi_n^{\dagger} \psi_0 \exp(i\omega_{n0}t)$$
$$\mathbf{j}_{n0} = -\psi_n^{\dagger} c\boldsymbol{\alpha} \psi_0 \exp(i\omega_{n0}t). \tag{5.30}$$

Then we integrate the term proportional to $\frac{\partial}{\partial z}\psi_n^{\dagger}\alpha_z\psi_0$ by parts over the z-coordinate and obtain

$$a_{fi}(\mathbf{b}) = i\frac{c}{v}\int_{-\infty}^{+\infty} dt \exp(i\omega_{n0}t)\langle\psi_n|\frac{v}{c}\Phi + \alpha_z\left(\frac{\Phi}{\gamma^2} - i\frac{v}{\omega_{n0}}\frac{\partial\Phi}{\partial z}\right)|\psi_0\rangle. \tag{5.31}$$

Taking into account that $\frac{\partial\Phi}{\partial z} = -\frac{1}{v}\frac{\partial\Phi}{\partial t}$ and

$$\int_{-\infty}^{+\infty} dt \exp(i\omega_{n0}t)\frac{\partial\Phi}{\partial t} = -i\omega_{n0}\int_{-\infty}^{+\infty} dt \exp(i\omega_{n0}t)\Phi \tag{5.32}$$

we arrive at the transition amplitude

$$a_{fi}(\mathbf{b}) = i\int_{-\infty}^{+\infty} dt \exp(i\omega_{n0}t)\langle\psi_n|\Phi\left(1 - \frac{v}{c}\alpha_z\right)|\psi_0\rangle \tag{5.33}$$

which coincides with the most "popular" form of the first Born amplitude (see, e. g., [16]).

A question of principle importance concerns the (asymptotic) high-energy limit $(\gamma \to \infty)$ of the eikonal amplitude (5.28). Since $\Phi = \gamma Z_A f(s_{\perp}, \gamma|z - vt|)$ one can show that

$$\int_{-\infty}^{+\infty} dt \exp(i\omega_{n0}t)\Phi(t) = \frac{2Z_A}{v}\exp(i\omega_{n0}z/v)\int_{-\infty}^{+\infty} d\xi f(s_{\perp}, \xi)$$

$$= \exp(i\omega_{n0}z/v)G\left(s_{\perp}, \frac{\omega_{n0}}{\gamma v}\right), \tag{5.34}$$

where the explicit form of G at the moment is not important. Correspondingly,

$$\int_{-\infty}^{+\infty} dt \quad \Phi(t) = G(s_{\perp}, 0) \equiv G_0(s_{\perp}) \tag{5.35}$$

and we obtain that

$$
a_{fi}(\mathbf{b}) = i\frac{c}{v\gamma^2}\langle\psi_n|\exp(iG_0)\exp(i\omega_{n0}z/v)G\alpha_z|\psi_0\rangle
$$

$$
- i\frac{c}{\omega_{n0}}\langle\psi_n|\exp(iG_0)\exp(i\omega_{n0}z/v)((\nabla_\perp G)\cdot\boldsymbol{\alpha}_\perp)|\psi_0\rangle. \tag{5.36}
$$

At impact energies so high that the difference between $G = G(s_\perp, \frac{\omega_{n0}}{\gamma v})$ and G_0 essentially vanishes, we may replace in (5.36) G_0 by G. Now we integrate by parts in the second line of (5.36), use the continuity equation (5.29) and then again integrate by parts. This transforms (5.36) into

$$
a_{fi}(\mathbf{b}) = \langle\psi_n|\exp(i\omega_{n0}z/v)\exp(iG)\left(1 - \frac{v}{c}\alpha_z\right)|\psi_0\rangle
$$

$$
- \frac{c}{v\gamma^2}\langle\psi_n|\exp(i\omega_{n0}z/v)\exp(iG)(1 - iG)\alpha_z|\psi_0\rangle \tag{5.37}
$$

whose limit at $\gamma \to \infty$ is given by

$$
a_{fi}(\mathbf{b}) = \langle\psi_n|\exp(i\omega_{n0}z/v)\exp(iG_0)\left(1 - \frac{v}{c}\alpha_z\right)|\psi_0\rangle. \tag{5.38}
$$

The amplitude (5.38) coincides with the amplitude derived in the so-called light-cone approach (see [17], [2]). The light-cone approach is strictly valid at $\gamma \to \infty$ and in this limit enables one to solve the problem of electron loss (ionization) and excitation exactly. In the case of strong ion-atom interaction this exact solution does not coincide with the Born results [2].

Thus, in the limit $\gamma \gg 1$ the symmetric eikonal model yields an exact solution for the transition amplitude for projectile-electron excitation and loss.

Using (5.23), we obtain

$$
\int_{-\infty}^{+\infty} dt\,\exp(i\omega_{n0}t)\Phi(t) = \frac{2Z_A}{v}\exp(i\omega_{n0}z/v)\sum_j A_j K_0(s_\perp\Lambda_j), \tag{5.39}
$$

where K_0 is the modified Bessel function [6], $s_\perp = |\mathbf{s}_\perp|$ and

$$
\Lambda_j = \sqrt{\kappa_j^2 + \omega_{n0}^2/(\gamma^2 v^2)}. \tag{5.40}
$$

We note that it also follows from (5.39)

$$
\int_{-\infty}^{+\infty} dt\,\Phi(t) = \frac{2Z_A}{v}\sum_j A_j K_0(\kappa_j s_\perp). \tag{5.41}
$$

Taking equations (5.39) and (5.41) into account the transition amplitude becomes

$$a_{fi}(\mathbf{b}) = i\frac{2Z_A c}{v^2} \sum_j A_j \langle \psi_n | \exp\left(i\frac{\omega_{n0}z}{v}\right) \exp\left(i\frac{2Z_A}{v} \sum_{j'} A_{j'} K_0(\kappa_{j'}s_\perp)\right)$$

$$\times \left(\frac{\alpha_z}{\gamma^2} K_0(s_\perp \Lambda_j) - i\frac{v\Lambda_j}{\omega_{n0}s_\perp} K_1(s_\perp \Lambda_j) \quad \mathbf{s}_\perp \cdot \boldsymbol{\alpha}_\perp\right) | \psi_0 \rangle, \tag{5.42}$$

where K_1 is the modified Bessel function [6]. Figure 5.2 presents results for electron loss from $Au^{78+}(1s)$ projectiles in collisions with neutral Au atoms at impact energies 1–30 GeV/u. These results include the first Born and the eikonal calculations as well as experimental data from [18] and [19] on the electron loss from 10.8 GeV/u $Au^{78+}(1s)$ projectiles. Even at these very high energies, the eikonal and first-order results still differ by 15–35 %. The calculated cross sections disagree with experimental data of [18] and [19], which themselves do not agree with each other.

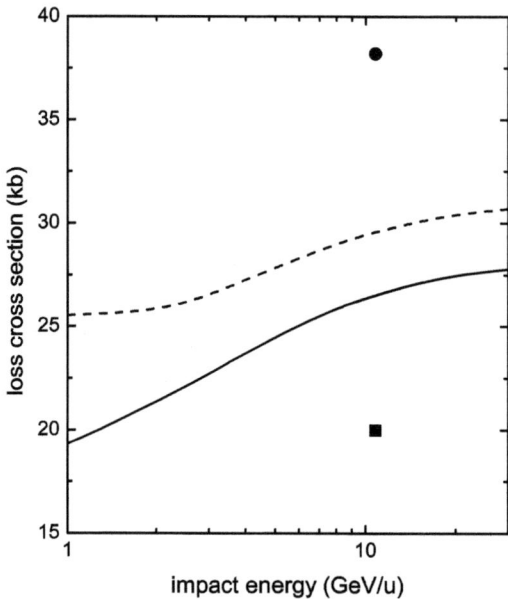

Figure 5.2: Cross section for the electron loss from $Au^{78+}(1s)$ ions in collisions with neutral Au atoms. Solid curve: eikonal results. Dash curve: first Born results. The circle displays experimental data from [18] while the square shows the result of [19] scaled to the gold target; from [12].

4 High-energy collisions with very light atoms I

Now we turn to projectile-electron transitions in high-energy collisions with a very light atom. In collisions with very light atoms, like e.g. hydrogen and helium, the contribution of atomic electrons to electron loss can be comparable to that of the atomic nucleus and, therefore, should be treated with especial care.

Unlike in the previous two sections, now the condition $Z_A/v \ll 1$ is always fulfilled. Concerning the ratio Z_I/v, we shall assume in this section that while the condition $Z_I/v \ll 1$ may be no longer met this ratio nevertheless remains substantially below 1,

$$\frac{Z_I}{v} < \sqrt{\frac{Z_A}{Z_I}}. \tag{5.43}$$

In such a case, the field of the ion may noticeably distort the motion of the atomic electron, which nevertheless is still mainly governed by the field of the atomic nucleus.

Under the condition (5.43), the first Born approximation may break down not because the atom itself represents a strong perturbation for the electron of the ion but because the distortion of the motion of the atomic electrons by field of the ion (not taken into account by the first Born approximation) affects in turn their interaction with the electron of the ion.

We again adopt the semiclassical treatment and assume that the atomic nucleus Z_A is at rest in our reference frame and taken as its origin. In this frame, the coordinates of the nucleus of the ion Z_I are given by $\mathbf{R}_I(t) = \mathbf{b} + \mathbf{v}t$.

For simplicity, we shall consider that the atom has only one (active) electron and assume for the moment that the process can be treated nonrelativistically ($v \ll c$ and $Z_I \ll c$). We denote the coordinates of the electron of the atom and that of the ion, given with respect to the atomic nucleus, by $\boldsymbol{\rho}$ and $\boldsymbol{\xi}$, respectively, and $\mathbf{s}$ and $\mathbf{r}$ are the coordinates of the atomic and ionic electrons with respect to the ionic nucleus.

Using the prior form (5.9) of the transition amplitude, we take $\Psi^{(-)}$ and χ_i according to [20]

$$\chi_i(t) = L_i u_0(\boldsymbol{\rho}) \exp(-i\epsilon_0 t) \psi_0(\boldsymbol{\xi} - \mathbf{R}(t)) \exp(i\mathbf{v} \cdot \boldsymbol{\xi}) \exp(-iv^2 t/2 - i\varepsilon_0 t)$$

$$\Psi^{(-)}(t) \approx \chi_f(t) = L_f u_m(\boldsymbol{\rho}) \exp(-i\epsilon_m t) \psi_n(\boldsymbol{\xi} - \mathbf{R}(t)) \exp(i\mathbf{v} \cdot \boldsymbol{\xi}) \exp(-iv^2 t/2 - i\varepsilon_n t). \tag{5.44}$$

The above states are represented by a product of an electron state of the ion and an electron state of the atom and also include distortion factors L_i and L_f for the atomic electron, which depend on $\mathbf{s}$. The states are not (anti) symmetrized assuming that the overlap between the electrons in the phase space is small.

Inserting (5.44) into the transition amplitude and keeping in mind that $n \neq 0$, we obtain

$$a_{fi} = a_{fi}^{ee} + a_{fi}^{eN}, \tag{5.45}$$

where

$$a_{fi}^{ee}(\mathbf{b}) = -i \int_{-\infty}^{+\infty} dt \exp(i(\varepsilon_n + \epsilon_m - \varepsilon_0 - \epsilon_0)t)$$

$$\times \int d^3\mathbf{r} \int d^3\boldsymbol{\rho} \, \psi_n^*(\mathbf{r}) u_m^*(\boldsymbol{\rho}) L_f^*(\mathbf{s}) \frac{1}{|\mathbf{R} + \mathbf{r} - \boldsymbol{\rho}|} L_i(\mathbf{s}) u_0(\boldsymbol{\rho}) \psi_0(\mathbf{r}) \tag{5.46}$$

and

$$
a_{fi}^{eN}(\mathbf{b}) = i \int_{-\infty}^{+\infty} dt \, \exp(i(\varepsilon_n + \varepsilon_m - \varepsilon_0 - \epsilon_0)t)
$$

$$
\times \int d^3\mathbf{r} \int d^3\boldsymbol{\rho}\, \psi_n^*(\mathbf{r}) u_m^*(\boldsymbol{\rho}) L_f^*(\mathbf{s}) \frac{Z_A}{|\mathbf{R}+\mathbf{r}|} L_i(\mathbf{s}) u_0(\boldsymbol{\rho}) \psi_0(\mathbf{r}). \tag{5.47}
$$

The term a_{fi}^{ee} describes transitions caused by the interaction between the electrons of the ion and atom. The other term, a_{fi}^{eN}, represents the contribution to the transition due to the interaction between the electron of the ion and the nucleus of the atom. For collisions, which are inelastic also for the atom ($m \neq 0$), this term would vanish if the distortion factors are replaced by 1 predicting (like the first Born consideration) that the simultaneous transitions in the projectile and target may be caused by the electron-electron interaction only. However, due to the presence of the **s**-dependent distortion factors, a_{fi}^{eN} becomes non-zero suggesting that the distorted-wave amplitude (5.45) contains yet another mechanism driving doubly inelastic collisions. Indeed, the term a_{fi}^{eN} describes simultaneous transitions in the ion and atom as caused by the joint effect of a single interaction between the electron of the ion and the atomic nucleus and multiple interactions of the atomic electron and the ionic nucleus. Thus, by the introduction of the distortion factors for the atomic electron one obtains the transition amplitude, in which not only the two-center electron-electron interaction but also the two-center electron-nucleus interactions are automatically taken into account in a relatively simple way.

It is of interest to note that the integrands in (5.46) and (5.47) do not contain derivatives of the distortion factors. Such derivatives contribute to the transition amplitude in the case when distorted-wave models are applied to ionization/excitation caused by collisions with structureless (or "frozen") particles (see, e. g., [3] and Section 3 of the present chapter). In the case under consideration, however, the terms containing the derivatives vanish because of orthogonality of the initial and final internal states of the structured projectile-ion.

Using equations (5.1) and (5.45), we obtain

$$
S_{fi}(\mathbf{q}_\perp) = -\frac{i}{4\pi^3 v} \int d^3\boldsymbol{\kappa}\, \frac{1}{\kappa^2} I_d(\mathbf{q}+\boldsymbol{\kappa}) I_p(\boldsymbol{\kappa}) I_t(\mathbf{q}, \boldsymbol{\kappa}), \tag{5.48}
$$

where

$$
I_d(\mathbf{p}) = \int d^3\mathbf{s}\, \exp(-i\mathbf{p}\cdot\mathbf{s}) L_f^*(\mathbf{s}) L_i(\mathbf{s})
$$

$$
I_p(\mathbf{p}) = \int d^3\mathbf{r}\, \psi_n^*(\mathbf{r}) \exp(i\mathbf{p}\cdot\mathbf{r}) \psi_0(\mathbf{r})
$$

$$
I_t(\mathbf{p}_1, \mathbf{p}_2) = \int d^3\boldsymbol{\rho}\, u_m^*(\boldsymbol{\rho}) \exp(i\mathbf{p}_1\cdot\boldsymbol{\rho})(1 - Z_A \exp(i\mathbf{p}_2\cdot\boldsymbol{\rho})) u_0(\boldsymbol{\rho}) \tag{5.49}
$$

and $\mathbf{q} = (\mathbf{q}_\perp; q_{min})$ with $q_{min} = (\varepsilon_n + \varepsilon_m - \varepsilon_0 - \epsilon_0)$.

In order to move further, one has to specify the form of the distortion factors. In the spirit of the symmetric eikonal model, we set [20], [2]

$$L_i(\mathbf{s}) = \exp(-iv_I \ln(vs + \mathbf{v} \cdot \mathbf{s}))$$

$$L_f(\mathbf{s}) = \exp(+iv_I \ln(vs - \mathbf{v} \cdot \mathbf{s})), \tag{5.50}$$

where $v_I = Z_I/v$. We note that the distortion factors (5.50) can be viewed as imposing the Coulomb boundary conditions on the initial and final states of the atomic electron.

With the distortion factors defined by (5.50), the transition amplitude (5.48) is given by

$$S_{fi}(\mathbf{q}_\perp) = -\frac{2i}{v^{1+2iv}} \int d^2\mathbf{p}_\perp f(\mathbf{p}_\perp, v_I) \langle \psi_n(\mathbf{r}) | \exp(i(\mathbf{p}_\perp - \mathbf{q}) \cdot \mathbf{r}) | \psi_0(\mathbf{r}) \rangle$$

$$\times \frac{1}{(\mathbf{q} - \mathbf{p}_\perp)^2} \langle u_m(\boldsymbol{\rho}) | Z_A \exp(i\mathbf{p}_\perp \cdot \boldsymbol{\rho}) - \exp(i\mathbf{q} \cdot \boldsymbol{\rho}) | u_0(\boldsymbol{\rho}) \rangle, \tag{5.51}$$

where

$$f(\mathbf{p}_\perp, v) = \lim_{\alpha \to +0} \lim_{\varsigma \to +0} \frac{\Gamma(1 - iv)\Gamma(1/2 + iv)}{2\pi\Gamma(1/2)\Gamma(2iv)} p_\perp^{\alpha - 2 + 2iv} \exp(-\varsigma p_\perp), \tag{5.52}$$

The derived expression (5.51) describes both the elastic ($m = 0$) and inelastic ($m \neq 0$) atomic modes but is only valid under the assumption that the initial and final internal states of the ion are different ($n \neq 0$).

The consideration presented in this section was purely nonrelativistic. However, the above eikonal model can also be generalized to the relativistic domain of the collision parameters [21] (see also [2]). The application of the nonrelativistic and relativistic eikonal models is illustrated in Figure 5.3. In particular, it follows from this figure that, rather unexpectedly, the models predict a noticeable deviation from the first Born results even when Z_I/v is as small as 0.1.

5 Collisions with very light atoms II

An electron in a highly charged ion is very tightly bound and, in order to remove it from the ion, the momentum transfers in the collision have to be sufficiently large. The minimum momentum transfers to the ion, q_{min}^I, and atom, q_{min}^A, are given by equations (5.2). Since the energy difference between the final and initial states of the ion $\varepsilon_f - \varepsilon_i$ is proportional to the square of the charge of the ion, it reaches large values in case of very highly charged ions that makes q_{min}^A quite big on the typical atomic scale even for large impact energies.

For example, let us suppose that $U^{91+}(1s)$ loses an electron in the collision with an atom at an impact energy of 1 GeV/u ($v \approx 120$ a. u., $\gamma \approx 2.07$). Applying equation (5.2) to

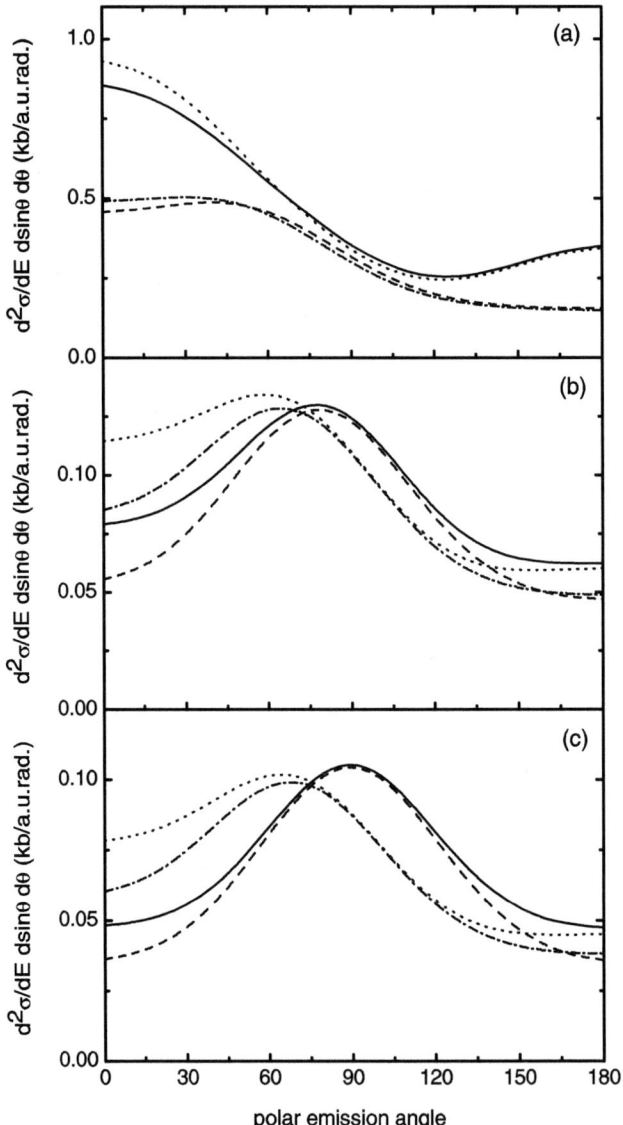

Figure 5.3: Angular distribution of 5 eV electrons emitted from helium in $O^{7+}(1s) + He(1s^2) \rightarrow O^{8+} +$ $He^+(1s) + 2e^-$ collisions. (a) 100 MeV/u (v = 58 a. u., $\gamma \simeq 1.1$); (b) 1 GeV/u (v = 120 a. u., $\gamma \simeq 2.1$); (c) 10 GeV/u (v = 136.5 a. u., $\gamma \simeq 11.7$); Solid curve: the relativistic eikonal calculation. Dash curve: the relativistic first-order result. Dot curve: the nonrelativistic ($c = \infty$) eikonal calculation. Dash-dot curve: the nonrelativistic first-order result; from [21].

this case, we obtain that q^I_{min} and q^A_{min} exceed 40 a. u. and 19 a. u., respectively. If the atom is very light, then q^A_{min} will be much larger than the typical orbiting momenta of all atomic electrons.

It is quite obvious that the increase in the values of the momentum transfers with Z_I goes together with the increase in the influence of the field of the highly charged ion on the motion of atomic electrons in the collision. The consideration of [22] shows that, provided the condition

$$\frac{Z_I}{\gamma v} \frac{Z_I}{Z_A} > \sqrt{\frac{Z_A}{Z_I}} \tag{5.53}$$

is fulfilled, the motion of the atomic electrons in the collision is governed by the field of the ion. As a result, the electrons and the nucleus of the atom act in the collision incoherently, the eikonal model discussed in the previous section becomes invalid and the impulse approximation represents a reasonable alternative leading to a conclusion that the cross sections for transitions of the electron of the ion can be written as

$$\sigma = Z_A^2 \sigma_p + Z_A \sigma_e. \tag{5.54}$$

Here, σ_p and σ_e are the contributions to the cross sections arising from collisions with protons and electrons, respectively.

One should add that, as "practical" calculations suggest, the condition (5.53) seems to be too strict and the impulse approximation can be used already under (much) softer condition

$$\frac{Z_I}{\gamma v} > \sqrt{\frac{Z_A}{Z_I}}. \tag{5.55}$$

Within the impulse approximation the state of an electron, which was before the collision bound in the atom (and which is necessary for calculating the contribution σ_e to the cross-section), in the rest frame of the ion can be approximated by [23]

$$\psi_i^{(+)}(\mathbf{r}, t) = \frac{\exp(-i\epsilon_a t/\gamma)}{\gamma} \int d^3\mathbf{p} \exp(-i\mathbf{p} \cdot \mathbf{R})$$
$$\times C_a\left(\mathbf{p}_\perp, -\frac{v\epsilon_a}{c^2} + \frac{p_z}{\gamma}\right) \chi_{\mathbf{p}}^{(+)}(\mathbf{r}), \tag{5.56}$$

Here, ϵ_a is the electron energy in the initial atomic state, $\chi_{\mathbf{p}}^{(+)}(\mathbf{r})$ is the Coulomb continuum "in-state" of the electron in the field of the highly charged nucleus with momentum $\mathbf{p} = (\mathbf{p}_\perp, p_z)$ and

$$C_a(\mathbf{k}) = \frac{1}{(2\pi)^{3/2}} \int d^3\mathbf{r}' \phi_0(\mathbf{r}') \exp(-i\mathbf{k} \cdot \mathbf{r}') \tag{5.57}$$

is the Fourier transform of the initial (nonrelativistic) atomic state ϕ_0. According to (5.56)–(5.57), the initial state of the electron is represented by a superposition of continuum states of the electron in the field of the ionic nucleus where the coefficients in this superposition are related to the momentum decomposition of the initial atomic state. We note that various modifications of the impulse approximation have been successfully used for considering radiative and nonradiative capture in high-energy asymmetric ion-atom collisions (see, e. g., [24], [25], [26], [27], [28] and references therein).

In calculating the cross section σ_e the state of the scattered electron is taken as an "out-state" of an electron moving in the field of the highly charged ion with an asymptotic momentum $\mathbf{p}_s$. We also note that since in collisions with large momentum transfers there may be a substantial overlap between the phase space of the electrons, initially bound in the ion and the atom, both the direct and exchange contributions to the transition amplitude have to be taken into account [22], [29].

Equation (5.54) was applied to excitation of 212.9 MeV/u U^{91+}(1s) in collisions with hydrogen and a very good agreement with experiment was found [30]. Equation (5.54) was also applied to electron loss from U^{91+}(1s) in collisions with hydrogen and helium [29]. However, available experimental data is not sufficiently accurate to test the theory. Figure 5.4 shows results for electron loss from highly charged ions in collisions with electrons, protons, and hydrogen and helium atoms calculated using the impulse approximation [29].

6 Conclusions

We have discussed a distorted-wave approach to the process of electron transitions in a highly charged ion (projectile-electron excitation or loss) caused by high-energy collisions with atoms. This approach can become very useful when one of (or both) the conditions $Z_I/v \ll 1$ and $Z_A/v \ll 1$ is (are) violated and, correspondingly, the Born approximation becomes invalid.

In collisions with many-electron atoms the field of the atomic nucleus can become too strong to be treated within the first Born approximation. In collisions with very light atoms, the field of the atom itself represents just a weak perturbation for the electron of the ion. However, the field of the ion can noticeably or even strongly influence the motion of atomic electrons that in turn affects their interaction with the electron of the ion. Since in collisions with very light atoms the contribution to electron transitions in the ion from the interaction with atomic electrons and nucleus are comparable in magnitude, the above point becomes quite important and has to be properly addressed.

For electron transitions in highly charged ions occurring in collisions with many-electron atoms at not very high impact energies the screening effect of atomic electrons can be very weak. In this range of the impact parameters, which can be characterized by the condition (5.4), one can neglect the presence of atomic electrons and treat electronic transitions in a highly charged ion as a three-body Coulomb problem which

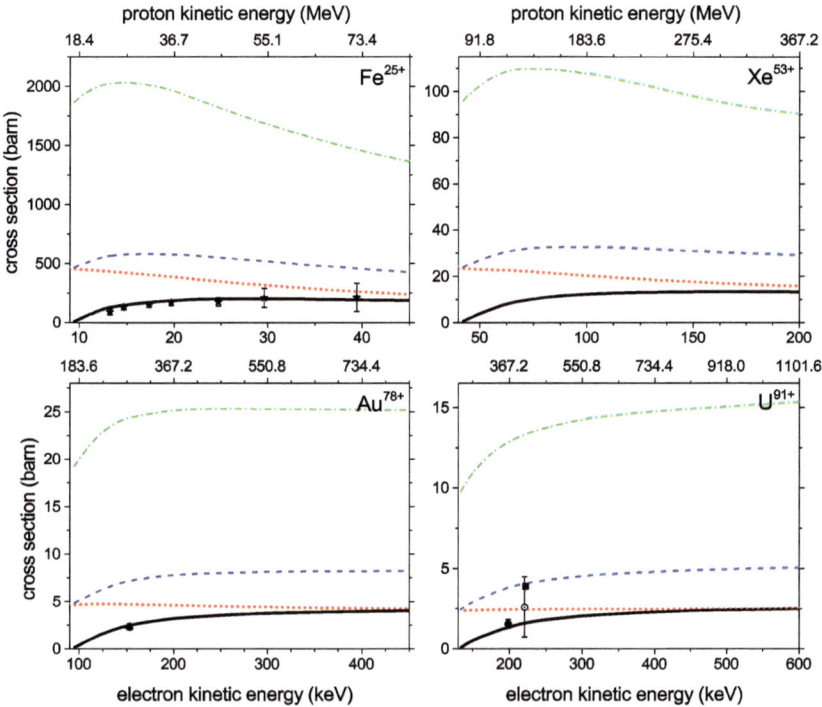

Figure 5.4: The total cross-sections for electron loss from $Fe^{25+}(1s)$, $Xe^{53+}(1s)$, $Au^{78+}(1s)$, and $U^{91+}(1s)$ in collisions with equivelocity electrons (solid curves), protons (dotted curves) as well as with atoms of hydrogen (dashed curves) and helium (dash-dot curves). In the right lower panel, the open circle with error bars is the total cross-section for electron loss from 405 MeV/u $U^{90}(1s^2)$ colliding with H_2 measured in [31] (which we scaled to collisions of hydrogen-like uranium with atomic hydrogen by diving their result by 4). All the other experimental data, shown in this figure, were measured for electron-ion collisions and are taken from [32] ($Fe^{25+}(1s)$), [33] ($Au^{78+}(1s)$), [34] ($U^{91+}(1s)$), [35] ($U^{91+}(1s)$). (From [29]).

can be dealt with by employing three-body Coulomb distorted-wave models like, e. g., continuum-distorted-wave–eikonal-initial-state and symmetric eikonal.

With increase in the impact energy larger and larger impact parameters begin to contribute to projectile-electron transitions. Therefore, in collisions with many-electron atoms at higher impact energies the presence of atomic electrons can no longer be neglected. In this range of the collision parameters, given by (5.15) or (5.16), the process of projectile-electron loss (or excitation) can be described by a symmetric eikonal model within the "frozen" atom approximation which assumes that the shape of the electron cloud in the atom remains unchanged during the very short collision time.

In collisions with very light atoms in the range of the impact parameters, where $Z_A/v \ll 1$ and the condition (5.43) is fulfilled, the effect of the field of the ion on atomic electrons can be accounted for by introducing distortion factors. The simplest distorted-wave model for such collisions is based on the symmetric eikonal approximation. Surprisingly, this model predicts noticeable deviations from the first Born results even when the ratio Z_I/v is still much less than 1.

With the increase in Z_I/v one eventually reaches the range of the collision parameters, given by (5.55), in which the field of the ion becomes during the collision the main driving field not only for the electron of the ion but also for those of the light atom. In this case, a proper description of projectile-electron transitions in the collisions can be made by employing a model based on the impulse approximation.

Bibliography

[1] McDowell MRC, Coleman JP. Introduction to the Theory of Ion-Atom Collisions. Amsterdam–London: North-Holland Publishing Company; 1970.

[2] Voitkiv AB, Ullrich J. Relativistic Collisions of Structured Atomic Particles. Berlin: Springer; 2008.

[3] Crothers DSF. Relativistic Heavy-Particle Collision Theory. London: Kluwer Academic/Plenum Publishers; 2000.

[4] Eichler J, Meyerhof W. Relativistic Atomic Collisions. San Diego: Academic Press; 1995.

[5] Crothers DSF, McCann J. J. Phys. B. 16 3229 (1983).

[6] Abramowitz M, Stegun I. Handbook of Mathematical Functions. New York: Dover Publications, Inc.; 1965.

[7] Maidagan JM, Rivarola RD. J. Phys. B. 1984;17:2477.

[8] Anholt R, Meyerhof WE, Xu X-Y, Gould H, Feinberg B, McDonald RJ, Wegner HE, Thieberger P. Phys. Rev. A. 1987;36:1586; Meyerhof WE, Anholt R, Xu X-Y, Gould H, Feinberg B, McDonald RJ, Wegner HE, Thieberger P. NIM A. 1987;262:10.

[9] Voitkiv AB, Najjari B. J. Phys. B. 2007;40:3295.

[10] Stolterfoht N, DuBois RD, Rivarola RD. Electron Emission in Heavy Ion-Atom Collisions. Springer; 1997.

[11] McGuire JH, Electron Correlation Dynamics in Atomic Collisions. Cambridge University Press; 1997.

[12] Voitkiv AB, Najjari B, Shevelko VP. Phys. Rev. A. 2010;82:022707.

[13] Greiner W. Relativistic Quantum Mechanics. 3rd ed. Berlin, Heidelberg, New York: Springer; 2000.

[14] Moliere G. Naturforsch. 1947;2A:133.

[15] Salvat F, Martinez JD, Mayol R, Parellada J. Phys. Rev. A. 1987;36:467.

[16] Eichler J. Lectures On Ion-Atom Collisions: From Nonrelativsitic To Relativistic Velocities. Amsterdam: Elsevier; 2005.

[17] Baltz AJ. Phys. Rev. Lett. 1997;78:1231.

[18] Westphal A, He YD. Phys. Rev. Lett. A. 1993;71:1160.

[19] Claytor N, Belkacem A, Dinneen T, Feinberg B, Gould H. Phys. Rev. A. 1997;55:R842.

[20] Voitkiv AB, Najjari B. J. Phys. B. 2005;38:3587.

[21] Voitkiv AB. Phys. Rev. A. 2005;72:062705.

[22] Najjari B, Voitkiv AB. Phys. Rev. A. 2012;85:052712.

[23] Najjari B, Voitkiv AB, Artemyev A, Surzhykov A. Phys. Rev. A. 2009;80:012701.

[24] Jakubassa-Amundsen DH, Amundsen PA. Z. Physik A. 1980;298:13.

[25] Dewangan DP, Eichler J. Phys. Rep. 1994;247:59.

[26] Belkic Dz, Comp J. Meth. Sci. and Tech. 2001;1:1.

[27] Hillenbrand P-M, Hagmann S, Atanasov D, Banas D, et al. Phys. Rev. A. 2014;90:022707.

[28] Hillenbrand P-M, Hagmann S, Jakubassa-Amundsen DH, Monti JM, et al. Phys. Rev. A. 2015;91:022705.

[29] Lyaschenko KN, Andreev OYu, Voitkiv AB. J. Phys. B. 2018;51:055204.

[30] Gumberidze A, Thorn DB, Fontes CJ, Najjari B, et al. Phys. Rev. Lett. 2013;110:213201.

[31] Hülskötter HP, Feinberg B, Meyerhof WE et al. Phys. Rev. A. 1991;44:1712.

[32] O'Rourke B, Currell FJ, Kuramoto H, Li YM, Ohtani S, Tong XM, Watanabe H. J. Phys. B. 2001;34:4003.

[33] Marrs RE, Elliott SR, Scofield JH. Phys. Rev. A. 1997;56:1338.

[34] Claytor N, Feinberg B, Gould H et al. Phys. Rev. Lett. 1988;61:2081.

[35] Marrs RE, Elliott SR, Knapp DA. Phys. Rev. Lett. 1994;72:4082.

Daniel Fischer

6 Recoil ion momentum spectroscopy with laser-cooled targets

1 Introduction

Understanding systems of interacting particles is one of the key challenges of physics, and has both fundamental and technological relevance. Such systems generally cannot be fully described in closed analytical form for more than two particles, even if their individual properties and the forces between them are precisely known. This dilemma is well known as the "few-body problem" and it limits the extent to which one can predict the state of the particles (classically it is given by their positions and velocities) for any time in the future. Therefore, advancing the knowledge of phenomena that emerge due to the complex interplay of several particles requires the joined theoretical and experimental exploration of a wide range of situations.

In the macroscopic world, the experimental study of few-body dynamics is straightforward: The positions and velocities of all interacting objects should be recorded as a function of time. This can, for instance, be done with a camera measuring time and location of the objects and deriving their velocities by comparing two subsequent frames. In this scenario, light is continuously scattered off the bodies and collected in the measuring device to learn about the bodies' motion. In the microscopic quantum-world of atoms and their constituents, such an approach has to fail. It is well known that a quantum measurement influences the system under investigation inevitably. For instance, the illumination of a bound electron with photons or other radiation that would allow resolving the electrons position with an uncertainty much smaller the size of the electron's initial wave function would require very short wavelengths of the scattered field which, in turn, would result in the transfer of large amounts of energy and momentum to the atomic particles. Therefore, any experimental approach to study few-body quantum-dynamics has to leave the system "un-observed" for sufficiently long time, before a measurement determines the values of specific observables in the "final" state. Generally, two steps are required: First, the system has to be prepared in a well-defined initial state controlling all (or at least most) of its degrees of freedom. Second, the final state has to be measured after a scattering reaction, i. e., the system has to be analyzed after its response to (internal or external) interactions. Due to the probabilistic character of the final state wave function, these two steps have to be repeated over and over again in order to obtain the probability distributions of the observables which correspond to the absolute square of the final state wave function.

Daniel Fischer, Missouri University of Science & Technology, Physics Dept., 1315 N. Pine St., Rolla, MO 65409, USA, e-mail: fischerda@mst.edu

https://doi.org/10.1515/9783110580297-006

In the last 3 decades, two complementary experimental techniques were developed in atomic, molecular, and optical physics independently of each other, which are ideally suited to perform the two above mentioned steps: First, there are laser-cooling and manipulation techniques [1, 2] which offer numerous possibilities to control few and many atom systems with highest fidelity. Magneto-optical traps (MOT), for instance, make it possible to trap gaseous atomic samples in a small volume well isolated from their environment and cool them down to temperatures below 1 mK. Second, with the advent of cold target recoil ion momentum spectroscopy (COLTRIMS), often referred to as "reaction microscopes" [3, 4], new and before inconceivable possibilities to study atomic fragmentation dynamics became accessible. This technique enables the mapping of the complete final state momentum space for up to 20 particles in coincidence and with high resolution. The combination of these two techniques—called MOTRIMS (magneto-optical trap recoil ion momentum spectroscopy) or MOTReMi (magneto optical trap reaction microscope)—represents one of the most advanced techniques to study atomic break-up dynamics.

The merging of laser-cooling with momentum spectroscopy is for several reasons very appealing: In comparison to traditional COLTRIMS experiments, where the target is provided by a gas-jet cooled through supersonic expansion, the atoms are typically by several orders of magnitude colder and the achievable momentum resolution is not limited by the target temperature. Moreover, in MOTs the targets can be prepared in excited and even polarized states, i. e., internal atomic degrees of freedom are accessible. The available atomic species are complementary to the atoms attainable by conventional targets. Low-temperature gas jets are limited to ground state noble gas atoms or molecular gases while MOTs allow to trap alkali and alkaline-earth metals which have a very different electronic structure with only one (or two, respectively) valence electron(s) and, therefore, allow to test dynamics in effective one- (or two-)electron systems.

However, the combination of laser-cooling and momentum spectroscopy bears also substantial experimental challenges. As a result, the overall number of MOTRIMS setups operated world-wide is much smaller than the one of conventional COLTRIMS experiments (it is presently below 10). An obstacle is connected to the different magnetic field configurations used for magneto-optical traps and COLTRIMS. Due to the quadrupole magnetic field employed in MOTs, it is generally not possible to unambiguously reconstruct the momenta of electrons emitted from a MOT target. Therefore, most of the MOTRIMS experiments focused on processes without free electrons in the final state ([5] and references therein) or where the electron momentum can be obtained indirectly from momentum conservation (e. g., [6]). To date, there is only a single experimental setup [7] that successfully combined coincident and momentum resolved electron and ion detection in COLTRIMS with a MOT target. With this experiment, it became possible to investigate all types of ionization reactions in full detail [8].

In this book chapter, the techniques of COLTRIMS and MOT as well as the combination of both are discussed and a selection of experimental results obtained with MOTRIMS are highlighted. While MOTRIMS has been applied to study many different processes including for instance (multi)photon ionization [6, 9] or even nuclear decay (e. g., [10]), in the present chapter only processes occurring in ion-atom collisions are considered focusing on electron transfer (see Section 4) as well as single ionization of lithium (Sections 5 and 6). Furthermore, this chapter does not to give a complete review, but it is rather meant to give a brief and selective introduction into the fields of momentum spectroscopy, laser-cooling, and ion atom collisions.

2 Momentum spectroscopy

Information on the dynamics of a few-body quantum-system can be obtained by measuring the bodies' momentum vectors or, equivalently, their angles and energies. The comparison of the measured distributions with theoretical differential cross-sections provides detailed insights into the effects and mechanisms at play.

In general, the experimentally observable quantities are governed by momentum and energy conservation which in non-relativistic approximation for ionizing atomic collisions in the frame where the target is initially at rest read as

$$\vec{p}_{\text{proj},i} = \vec{p}_{\text{proj},f} + \vec{p}_{\text{tar}} + \sum_n \vec{p}_{\text{elec},n} \tag{6.1}$$

$$K_{\text{proj},i} + E_{\text{bind},i} = K_{\text{proj},f} + E_{\text{bind},f} + K_{\text{tar}} + \sum_n K_{\text{elec},n} \tag{6.2}$$

with $\vec{p}$ denoting the momenta of projectile, residual target ion, and emitted electrons in the initial and final state, and K and E_{bind} being the kinetic and electronic binding energies, respectively. Experiments in which all the magnitudes shown in the above equations are determined are called "kinematically complete".[1] In this section, an overview of the experimental techniques allowing to measure momenta of atomic particles is given.

2.1 Milestones of differential scattering measurements

The well-known series of scattering experiment by H. Geiger, E. Marsden, and E. Rutherford performed from 1908 to 1910 [11, 12, 13], where α-particles were scattered

---

[1] The complement "kinematical" is used because a truly "complete experiment" would include the information on the angular momentum balance. In practice, complete experiments are not easily performed because with present day techniques it is not possible to measure the particles' spins in coincidence.

off a very thin gold foil, can be regarded as the genesis of atomic momentum spectroscopy. All later differential experiments rely at least partly on the same concept than those performed in Rutherford's laboratory: The initial state of the particles was prepared with sufficiently high accuracy with respect to their energies, momenta, and interaction coordinate. After the collisions, the particles' positions were measured far away from the reaction volume which revealed information on the scattering direction. However, if more than two particles are involved or if the collision is inelastic Rutherford's method generally constitutes an incomplete measurement because not all kinematic quantities in the final state (see equations (6.1) and (6.2)) are determined unambiguously. In general, there are two more requirements for a complete measurement: First, several particles—for instance the scattered projectile as well as at least one of the target fragments—have to be detected in coincidence and, second, the particles' energies need to be recorded and not only their emission angles.

With the advent of coincidence measurements—first performed by W. Bothe and H. Geiger [14] for Compton scattering in 1924—and high-resolution electron spectroscopy [15, 16, 17] the way was paved for kinematically complete measurements on a larger variety of collision systems. A significant milestone was achieved with the first kinematically complete experiment on electron-impact ionization—the so-called (e, 2e) experiment[2]—which was reported by H. Ehrhardt et al. in 1969 [18]. In this experiment, helium atoms were bombarded with a mono-energetic electron beam and two particle analyzers with small entrance slits were employed to observe the scattered projectile and the emitted electron in coincidence. The scattering and emission angles were determined geometrically by the position of the entrance slit of the analyzers and the energy of one of the electrons was measured in an electrostatic spectrometer (see Figure 6.1). The energy of the second electron and the momentum transfer vector to the remaining target ion was determined by energy and momentum conservation, respectively. In this way, fully differential cross-sections $d^3\sigma/dE_e d\Omega_e d\Omega_s$ (see note in Chapter 2 of this book) became accessible.

The general technique of energy and angle dispersive electron analyzers was for several decades the "gold standard" in differential scattering experiments and is used in many laboratories until today. However, this type of measurement is afflicted with some substantial limitations: First, the spectrometers' energy and/or angular acceptance is very limited and measuring a distribution over a larger range requires scanning the spectrometer voltages or the physical position of the whole analyzers, respectively. This results in very long measuring times and complicated cross-normalization procedures of the data taken for different energies and angles. Second, while this technique works well for electrons it does generally not work for heavier particles such as

---

2 "(e, 2e)" reflects the number of unbound electrons in the initial and final state, respectively. The term was coined in analogy the (p, 2p) reaction from nuclear physics which involves colliding nuclei with protons and detecting the scattered and the emitted protons in coincidence.

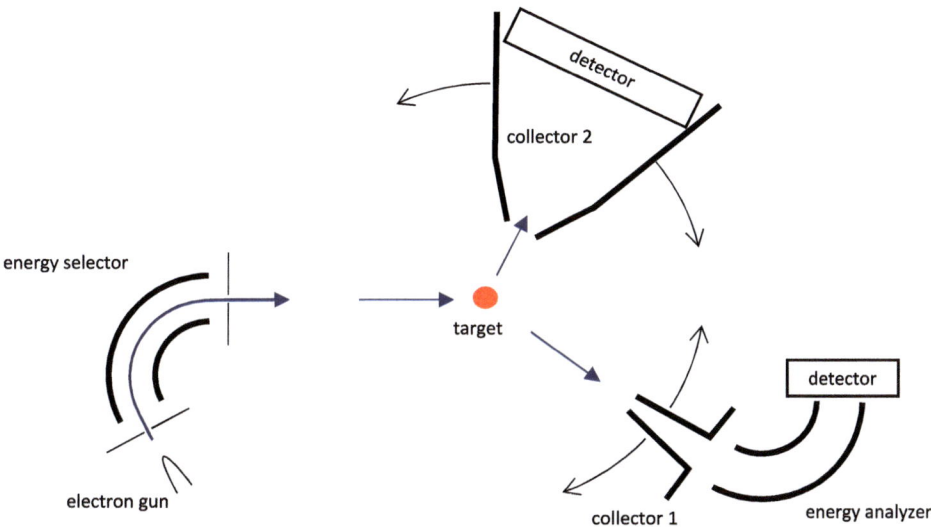

Figure 6.1: Schematic setup of the first (e, 2e) experiment [18]. The scattered and emitted electrons are measured in collectors 1 and 2, respectively.

ions. There are only very few experiments reported in literature with moderate collision energies (up to 200 keV proton beams, e. g., [19]) where ionic projectiles are measured with an angular and energy dispersive spectrometer yielding sufficient resolution to obtain high quality data. For higher projectile energies or masses, the deflection angle (typically in the range 10^{-5} rad or below) and the relative energy loss (10^{-5} to 10^{-9}) of the projectiles in collisions is so small that it cannot easily be measured directly and often is completely blurred by the projectile beam emittance.

In the 1980s, an alternative method has been developed: Recoil ion momentum spectroscopy (RIMS). The idea of this approach is to measure the recoil momentum of the target atom that is ionized in a collision instead of the projectile. To this end, the ionized target is measured with a large position and time sensitive detector and its momentum is calculated by the recorded time and position information. This technique first implemented by J. Ullrich and H. Schmidt-Böcking [20] on 5.9 MeV/u U^{65+} on Ne collisions was found to provide far superior resolution than any other method aiming to measure the scattering angle of the heavy projectile ions directly. However, the achieved momentum resolution was still not competitive with electron momentum measurements mainly due to the thermal motion of the gas atoms which is much more critical for heavy particles than for electrons. This can be easily seen from the differential representation of the nonrelativistic energy-momentum relation $\vec{p} \cdot \mathrm{d}\vec{p} = m\mathrm{d}E$. Let us assume, a proton and an electron with identical momentum $\vec{p}$ shall be detected with the same momentum resolution $\mathrm{d}\vec{p}$. In this case, the energy uncertainty $\mathrm{d}E$ of the proton measurement needs to be m_p/m_e (=1836) times lower than the one for the electron. A moderate electron energy resolution of 1 eV corresponds to a resolution of well below 1 meV for the proton. For a target gas at room temperature with a mean kinetic energy of about 30 to 40 meV, such a resolution is unequivocally unachievable.

The breakthrough came with two further advancements of the RIMS technique: On the one hand, supersonic gas jets were introduced to create well collimated target beams with internal temperatures below 1 K corresponding to thermal energies of just few ten μeV (e. g., [21, 22]). This improvement put the momentum resolution of RIMS on a par with the resolution of traditional electron spectrometers. On the other hand, the recoil ion momentum spectrometer was enhanced with a second detector for electrons [23]. By combining electric and magnetic fields, both recoil ions and electrons could be guided to the two detectors and be recorded in coincidence with high momentum resolution over a solid angle of nearly 4π. The new type of spectrometers was interchangeably called cold target recoil ion momentum spectroscopy (COLTRIMS) or reaction microscope. This novel technique enabled kinematically complete experiments for all types of projectiles because the energy and momentum change of the projectile can always be calculated from the conservation laws (equations (6.1) and (6.2)). The first fully differential cross-sections for ion impact were reported 2003 [24]. Nowadays, COLTRIMS is one of the standard tools whenever the dynamics in atomic scattering and fragmentation processes is investigated.

2.2 COLTRIMS

COLTRIMS spectrometers are operated in more than hundred laboratories worldwide and, obviously, there is a very large variety of designs and momentum imaging schemes which are adapted for very specific experimental situations. Key technologies that are used in COLTRIMS are large-area microchannel plate (MCP) detectors with position sensitive read-out as well as the cooling of target gases in supersonic gas-jets. While these particular techniques represent a significant contribution and are fields of very active research and development until today, their detailed discussion will be omitted here and the reader is referred to [3, 4, 25]. The aim of this section is to give a general overview of the working principle of momentum imaging and discuss some general features and design considerations that have proven to be useful.

General working principle of COLTRIMS
Before discussing the general scheme of COLTRIMS, we first consider a situation which similarly can be found in many classical mechanics textbooks: A person throws a small object from a tower and, due to gravity, the object travels on a parabolic trajectory and, eventually, hits the ground. Analyzing the trajectory using for instance the Newtonian approach makes immediately clear that the position where it hits the ground and the time it travels through the air depend on magnitude and direction of the object's velocity as it leaves the hand of the thrower. Moreover, the interrelation

of travel time and impact position with the starting velocity is unambiguous, i. e., the information on flight duration and relative position of start and end point is sufficient to derive the object's initial velocity vector.

In a similar way, the momentum of charged target fragments emerging from an ionizing atomic collision can be obtained in a COLTRIMS experiment. In the center of the spectrometer is the reaction volume (typ. $\leq 1\,\mathrm{mm}$ of diameter) where a cold gaseous target is overlapped with a projectile beam consisting of ions, electrons, photons, antiprotons, or positrons. When a target atom is ionized, the emitted electron(s) and the recoiling target ion fly away from the reaction location with a "starting" momentum that is to be determined. With an electric field, electrons and recoil ions are extracted in opposite directions and deflected onto two detectors located on opposing sides of the spectrometer. An additional magnetic field is used to confine the motion of the particles, thereby effectively increasing the collection efficiency of the setup. The detectors record all the fragments' splat positions and times in coincidence which depend on the "starting" momenta of the particles.

While the situation in COLTRIMS has many analogies to the classical example discussed above, there are two slight complications to be considered: First, the flight duration of the particles in the spectrometer is not directly accessible but only the time of arrival at the detectors. Therefore, a reference time is required which is synchronized to the instant of the actual scattering process. This reference time is usually obtained by using short-pulsed projectile beams or, in some cases, by the coincidence signal of the scattered projectile. Second, the configuration of the electric and magnetic field has to allow for an unambiguous reconstruction of the particles' momenta. In the classical example, the rock is deflected toward the detector (i. e., the ground) by the earth's gravitational field which is conservative and, in very good approximation, homogenous in the boundaries of the "spectrometer." In COLTRIMS, the electric or magnetic fields are generally not so well behaved, because the electric field is not necessarily constant throughout the spectrometer range and the magnetic field introduces a nonconservative force. On the downside, this can result in ambiguities which limit the resolution or hamper the reconstruction for certain momentum regions. On the positive side, the additional flexibility in the choice of the fields can be exploited to enhance the energy acceptance and improve the resolution by employing focusing techniques.

Charged particles in static parallel electric and magnetic fields

In order to analyze the target fragments trajectories in the momentum spectrometer, we start with the most fundamental case of a particle of charge q and mass m traveling in static electric and magnetic fields $\vec{E}(\vec{r})$ and $\vec{B}(\vec{r})$. The force on the particle in the nonrelativistic approximation is given by the Lorentz Force Law:

$$\vec{F}(\vec{r}, \vec{v}) = m\frac{\mathrm{d}^2}{\mathrm{d}t^2}\vec{r} = q(\vec{E}(\vec{r}) + \vec{v} \times \vec{B}(\vec{r})). \tag{6.3}$$

The simplest configuration that is used in COLTRIMS is with both fields, electric and magnetic, being homogenous and oriented in the z-direction, i. e., $\vec{E}(\vec{r}) = (0,0,E_z)$ and $\vec{B}(\vec{r}) = (0,0,B_z)$. In Cartesian coordinates, the above equation simplifies for these fields to

$$m\frac{d^2}{dt^2}x = qB_z\frac{dy}{dt} \tag{6.4}$$

$$m\frac{d^2}{dt^2}y = -qB_z\frac{dx}{dt} \tag{6.5}$$

$$m\frac{d^2}{dt^2}z = qE_z. \tag{6.6}$$

As can be seen, the motion of the particle in the xy-plane, i. e., in the plane perpendicular to the fields, is governed by the magnetic field and the motion in the z-direction by the electric field. The whole set of differential equations can be solved for the initial conditions $\vec{r}(0) = 0$ and $\vec{p}(0) = (p_{0x}, p_{0y}, p_{0z})$ as follows:

$$x(t) = \frac{1}{qB_z}(\sin(\omega_c t)p_{0x} + (1 - \cos \omega_c t)p_{0y}) \tag{6.7}$$

$$y(t) = \frac{1}{qB_z}((\cos \omega_c t - 1)p_{0x} + \sin(\omega_c t)p_{0y}) \tag{6.8}$$

$$z(t) = \frac{p_{0z}}{m}t + \frac{qE_z}{2m}t^2 \tag{6.9}$$

with $\omega_c = \frac{qB_z}{m}$ being the cyclotron frequency of the particle. The trajectory calculated above describes a circle in the xy-plane with its center at $(x,y) = \frac{1}{qB_z}(p_{0y}, -p_{0x})$ and a radius of $\sqrt{p_{0x}^2 + p_{0y}^2}/qB_z = \frac{p_{0\perp}}{qB_z}$. In the z-direction, the particle undergoes a motion of constant acceleration. After the time-of-flight T, the particle reaches the detector which is placed at the position $z = z_d$. T can be derived from equation (6.9) to be

$$T = \frac{1}{qE_z}(\sqrt{p_{0z}^2 + 2z_d mqE_z} - p_{0z}). \tag{6.10}$$

The coordinates of the particle as it hits the detector are $x(T) = x_d$, $y(T) = y_d$, and $z(T) = z_d$ and are measured in the experiment (for the x and y directions) or known from the experimental geometry (for z), respectively. Using this information, equations (6.7)–(6.9) can be solved for the starting momentum of the particle:

$$p_{0x} = \frac{qB_z}{2}\left(\frac{\sin \omega_c T}{1 - \cos \omega_c T}x_d - y_d\right) = \frac{qB_z}{2}\left(\cot\left(\frac{\omega_c T}{2}\right)x_d - y_d\right) \tag{6.11}$$

$$p_{0y} = \frac{qB_z}{2}\left(x_d + \frac{\sin \omega_c T}{1 - \cos \omega_c T}y_d\right) = \frac{qB_z}{2}\left(x_d + \cot\left(\frac{\omega_c T}{2}\right)y_d\right) \tag{6.12}$$

$$p_{0z} = \frac{z_d m}{T} - \frac{qE_z}{2}T \tag{6.13}$$

Equations (6.11), (6.12), and (6.13) contain all of the math required to reconstruct the starting momentum of the particle from the recorded data.

Electrons and ions in the spectrometer field

So far, we did not distinguish between electrons and recoil ions and the above derived relations are valid for either particle. However, for given electric and magnetic fields the trajectories of ions and electrons can be very different from each other due to their large difference in mass. For analyzing the trajectories of the particles, the desirable field strengths should first be estimated. They have to be optimized for both resolution and acceptance in the desired momentum range. For the following discussion, we assume that the starting momenta of ions and electrons are similar in magnitude,[3] that the ions are singly charged, and that all particles with a momentum of up to p_{max} should be detected irrespective of their emission angle.

First, we analyze the motion of the particles in the direction parallel to the fields, i. e., the z-direction. We note that the strength of the electric field has to be chosen sufficiently high to allow for the collection of even those particles that have a starting momentum of p_{max} directed away from the detector. In other words, the maximum potential energy gained while traveling away from the detector $qE_z z_d$ should have the same magnitude as the maximum starting kinetic energy $K_{max} = p_{max}^2/2m$. Due to their lower mass, electron energies are higher by a factor of m_{tar}/m_e than those of the ions with the identical momentum. This results in a lower limit for the electric field $E_z = p_{max}^2/(2m_e e z_d)$ with e being the elementary charge. The analysis equation (6.10) for the starting momentum p_{max} shows now that the time T is almost six times larger for an electron moving away from the electron detector than for one moving toward it (see Figure 6.2). That is, the time of flight shows a very large relative dispersion with respect to the electrons' momenta in z-direction.

The situation is very different for the recoil ions exposed to the same electric field. For singly charged ions that move away or toward the ion detector with a momentum of p_{max} we obtain a relative variation in T of only about $\sqrt{m_e/m_{tar}}$ which is about 1.5 % for lithium and even smaller for heavier target atoms (inset in Figure 6.2). That means, the recoil ions time-of-flight distribution will feature a narrow peak and the starting momentum of the ion will affect the measured flight time only by a very small percentage. In most cases the ion z-momentum can even be calculated using the linear

---

3 For "soft" collisions, the momenta are typically in the range of a few atomic units, i. e., $\sim 10^{-25}$ to 10^{-23} kg m/s. However, there are exceptions. Recoil ion momenta can be significantly larger than electron momenta for instance if a molecular target breaks up in several ionic fragments. In this case, the Coulomb repulsion between the ions can result in kinetic energies of several eV. This situation is also referred to as "Coulomb explosion." Electron momenta can be much larger than ion momenta, too, e. g., for electrons emitted in "violent" ion-atom collisions or for projectile electrons in (e, 2e) experiments.

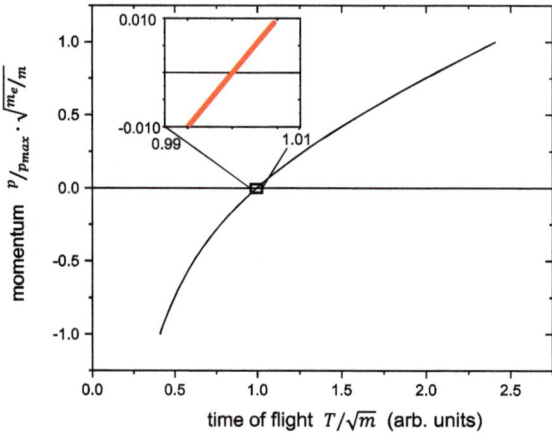

Figure 6.2: Relation between generalized time-of-flight and z-momentum of the particles according to equation (6.13). The black solid line and the red line in the inset represents the curves for electrons and ${}^6Li^+$ ions, respectively, with z-momenta between $-p_{max}$ and $+p_{max}$.

approximation of equation (6.10) for $p_{0z} \ll \sqrt{2z_d mqE_z}$, yielding

$$p_{tar\,z} \approx qE_z(T_0 - T) \tag{6.14}$$

with T_0 being the time-of-flight of ions with zero starting momentum.

In the direction perpendicular to the fields, the motion is governed by the magnetic field. As shown from equations (6.7) and (6.8) before, the particles undergo a circular motion with the radius $p_{0\perp}/qB_z$. For full detection efficiency of all particles with a momentum p_{max} or smaller, the particles distance to the z-axis should never exceed the radius R_d of the detector. Therefore, the radius of the gyration motion should not be larger than $R_d/2$. This leads to the requirement for the minimum magnetic field strength of $B_z = 2p_{max}/qR_d$. Now we compare again the behavior of electrons and ions with the same starting momentum in that field. While the radius of the gyration motion is identical for both, the angular frequency ω_c for the electrons is larger by a factor of m_{tar}/m_e.

Comparing the gyration period $T_c = 2\pi/\omega_c$ with the overall time-of-flight shows that an electron undergoes several full turns before it hits the detector. How often it cycles and at which gyration phase it arrives at the detector depends very much on its flight time T, i. e., on its starting z-momentum. While the magnetic field significantly enhances the collection efficiency of the electrons, the helical trajectory results also in ambiguities in the reconstruction of the x and y momentum components: If the time-of-flight of the electron is such that $T\omega_c$ is an integer multiple of 2π, equations (6.11) and (6.12) cannot be solved due to the poles of the cotangent. In this case, the electron will hit the detector at $x_d = y_d = 0$ no matter what the transverse starting momentum was (see Figure 6.3).

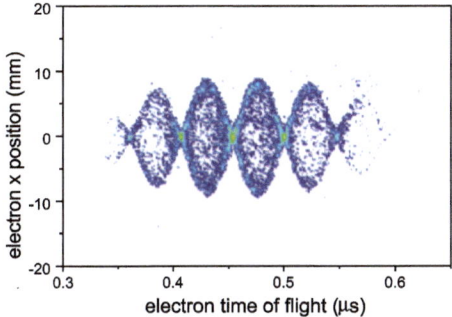

Figure 6.3: Electron yield as a function of their time-of-flight T and their x-position on the detector for photo-ionization of lithium. Due to energy conservation, all emitted electrons have a kinetic energy of 1.1 eV in this experiment. In accordance with equation (6.7), the distribution features nodes in x for flight times T being integer multiples of $2\pi/\omega_c$.

For the ions the reconstruction of the transverse momenta is significantly easier mainly due to two reasons: First, the gyration period is typically much larger than the time of flight, i.e., $\omega_c T \ll 2\pi$. Therefore, the ambiguities in the momentum reconstruction occurring for electrons are generally not relevant for ions. Second, due to the very small relative variation of the ion's time-of-flight the angle $\omega_c T$ can in most cases just be assumed to be a constant and does not need to be determined for each ion individually. One can conclude that the ions trajectories are much less affected by the magnetic field than the electrons.

Experimental resolution, focusing techniques, and spectrometer design
The idealized conditions in the above discussion are, of course, not fulfilled in a real-world experiment where uncertainties limit the achievable resolution. Commonly considered resolution-limiting factors are the temporal and spatial uncertainty of the detector measurements, the initial temperature of the target, electric and magnetic fringe fields, and the finite size of the reaction volume. The strategies to optimize the resolution is for most of these effects evident, e. g., tweaking the performance of detectors and electronics, cooling the target to lowest possible temperatures, and avoiding fringe fields by carefully considering spectrometer materials and electrode design. When designing a spectrometer, deciding on its overall size is a very important aspect, too, because for larger spectrometers the influence of detector resolution, fringe fields, and the size of the reaction region is reduced.

The influence of the size of the reaction volume can efficiently be minimized by employing focusing techniques with modified field configurations. In order to analyze the influence of the reaction range on the resolution, we first go back to the equations describing the particles motion (equations (6.11)–(6.13)) and vary slightly the initial condition for the particles' coordinate. For a simple quantitative estimate in z-direction, we consider a particle with zero momentum whose starting position is altered by $\Delta z \ll z_d$. In first order, equation (6.10) shows that the different starting position results in a small change in the time-of-flight with

$$\Delta T \approx \frac{\Delta z}{2z_d} T. \tag{6.15}$$

Using now the altered time-of-flight in the calculation of the starting z-momentum given in equation (6.13) yields a nonzero calculated momentum of

$$\Delta p_z = qE_z T \frac{\Delta z}{2z_d} = \sqrt{2mqE_z z_d} \frac{\Delta z}{2z_d}. \tag{6.16}$$

From this equation, it can already be seen that the influence of the altered starting point on the calculated momentum scales with the square root of the mass of the particle. However, the resolution also depends on the electric field strength. Using the considerations of the previous subsection and choosing the electric field such that electrons can be detected for all emission angles up to a momentum of p_{max} yields for singly charged particles a momentum uncertainty of

$$\Delta p_z = \sqrt{\frac{m}{m_e}} p_{max} \frac{\Delta z}{2z_d}. \tag{6.17}$$

While this equation holds for both electrons and ions, the quantitative result is very different for the two species. As an example, if the extension of the reaction volume is about 1 % of its distance to the detector the expected momentum resolution for electrons in z-direction is about 0.5 % of the maximum electron momentum. Turning now to the ions, the momentum uncertainty is by a factor of $\sqrt{m_{tar}/m_e}$ higher. For singly charged helium ions, e. g., the uncertainty would be almost 43 % of the maximum detectable electron momentum. This analysis shows: While electrons can still be detected with good resolution, a meaningful measurement of the recoil ion momentum is largely impossible.

This dilemma can be resolved by modifying the electric field and introducing a dependence on the z-coordinate $E_z = E_z(z)$.[4] There are field configurations such that the time-of-flight in first order does not show a dependence on a small variation of the starting point, i. e., $\partial T(\Delta z, p_{0z})/\partial \Delta z = 0$. This situation is referred as to "time focusing." The most commonly used time focusing configuration is having a constant electric field over a distance a followed by a field free drift region of length $d = 2a$ which is often referred to as Wiley–McLaren configuration [26]. Conceptually, the focusing effect can be understood by considering two particles with the same starting

---

4 Of course, the spatial variation of only one component of the electric field without affecting the other ones is impossible in a source-free region (i. e., $\rho(\vec{r}) = 0$) because of $\vec{\nabla} \cdot \vec{E} = 0$. However, this effect will be generously overlooked in the present discussion.

momentum but with different starting position. The particle further away from the detector travels a larger distance through the accelerating field, thereby gaining a higher speed. In the field-free drift region, it has time to catch up with the slower particle that has a smaller distance of travel but a lower speed. For the right length of the drift region, both particles have exactly the same travel time.

The uncertainties in the initial x and y-coordinates increase the experimental error, too. Due to the linearity of equations (6.11) and (6.12), an estimate on the relative error of the transverse starting momentum $p_{0\perp}$ is straightforward and it is

$$\frac{\Delta p_{0\perp}}{p_{0\perp}} = \frac{\Delta r_\perp}{r_\perp} \tag{6.18}$$

with $\Delta r_\perp = \Delta x = \Delta y$ being the diameter of the reaction volume and $r_\perp = \sqrt{x^2 + y^2}$ the distance from the hit position to the center of the detector. It should be noted that the uncertainty in the transverse direction is in general less severe than that for the longitudinal ion momentum as discussed above. For the recoil ions, this error can further be reduced, by employing an electrostatic lens close to (or, even better, right at) the reaction volume. The idea here is, to slightly accelerate the ions near their starting point toward the center of the detector. In this way, the finite size of the starting volume can efficiently be focused to a much smaller spot size on the detector while the position spread due to the starting momentum distribution remains largely unaffected. However, when employing such a focusing scheme three aspects should be considered: First, the spatial electric lens will affect the time-focusing. This can be compensated for by choosing a longer drift region. Second, the presence of a magnetic field will reduce the focusing effect and can even nullify it completely. This is because the initial velocity toward the detector center is redirected by the Lorentz force. Therefore, the angle $\omega_c T$ accumulated by the ion during its flight should be kept as small as possible. And third, the electrostatic lens for the ions will also affect the trajectory of electrons and can increases the uncertainty in the determination of their momentum.

In Figure 6.4, different field configurations for a COLTRIMS spectrometer are illustrated (from [7]). In this design, the electric field is generated by a large number of ring electrodes. This design provides a large flexibility and the field strength and its dependence on the z-coordinate can be easily modified by applying appropriate voltages to the electrodes.

3 MOTRIMS

The advent of laser technology in the 1960s extended significantly the experimental possibilities of analyzing, manipulating, and controlling atoms and their motion with tremendous precision. Among the large variety of techniques and schemes that were developed, the Magneto-Optical Trap (MOT), first realized in the 1980s [27], became

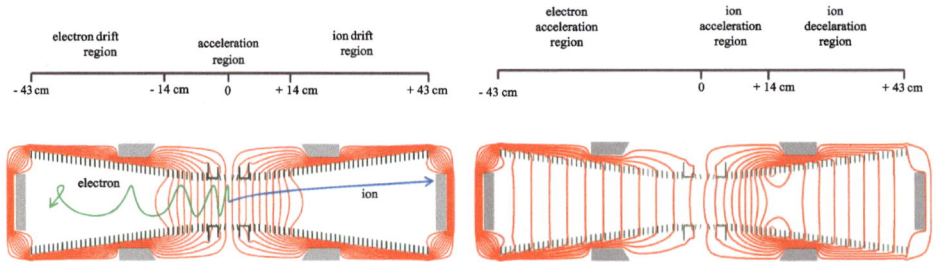

Figure 6.4: Electric field configurations of the momentum spectrometer from [7]. The red lines indicate traces of equal electro-static potential. Shown are configurations with time-focusing [26] for electrons and ions (left) and with position and time focusing for the ions (right).

one of the workhorses in the field, because it allows to trap large amounts of atoms (typ. 10^7 to 10^9 atoms) in a small volume (a few mm in diameter) and cool them to sub-mK temperature. That makes MOTs a very interesting option for COLTRIMS scattering experiments. The combination of the two techniques is dubbed MOTRIMS (magneto-optical trap recoil ion momentum spectroscopy) if only recoil ions are measured, or MOTReMi (magneto-optical trap Reaction Microscope) for coincident electron and ion detection. In this section, a brief overview of the relevant techniques is given and some of the technical advantages and challenges are discussed.

3.1 Laser-cooling and trapping of atoms

Laser-cooling and optical trapping is a very wide field and many techniques were invented in the last 30 years allowing to manipulate atoms, ions, molecules, or even larger objects such as living cells. There is a fair amount of literature describing and explaining the large variety of methods developed and phenomena observed. The aim of this section is to give a very brief introduction to some simple concepts and basic principles of Doppler-cooling and magneto-optical traps. While the discussed models are useful to understand qualitatively some general features of laser-cooled samples, the quantitative description of the properties of a trap, e. g., with respect to temperature, density, or population distribution requires more rigorous approaches. For a more complete introduction, the reader may be referred to [1, 2].

We start by considering a two-level atom which is exposed to a monochromatic directed laser beam of frequency ω and wave vector $\vec{k}$. If the atom absorbs and reemits photons at the rate Γ, it will experience a mean force due to the photon absorption of

$$\langle \vec{F} \rangle = \Gamma \hbar \vec{k} \tag{6.19}$$

The spontaneous decay of the atom back to the ground state will result in a vanishing average force after a large number absorption and reemission cycles, because the angular profile of the spontaneous reemission of the photons features point symmetry to the origin, i. e., it happens in any two opposing directions with the same probability. The force given in equation (6.19) is also referred to as "spontaneous" force. The photo-excitation of the atom is a resonant process and, therefore, the scattering rate Γ depends strongly on the frequency ω of the incoming laser radiation. The resonance exhibits the Lorentzian shape of a driven damped harmonic oscillator and it is given by

$$\Gamma(\omega) = \gamma \frac{I/2I_s}{1 + I/I_s + (2(\omega - \omega_0)/\gamma)^2} \tag{6.20}$$

with I being the intensity of the laser beam, and with ω_0, γ, and I_s the resonance frequency, natural line width, and saturation intensity, respectively, which are all properties of the atomic transition. Naturally, the highest scattering rate is expected for a laser frequency of $\omega = \omega_0$. Considering the velocity $\vec{v}$ of the atom with respect to the laser source, the Doppler effect results in an apparent shift of the laser frequency by $\vec{k} \cdot \vec{v}$. Putting the scattering rate of equation (6.20) into equation (6.19) and considering the Doppler shift results in a velocity and laser-frequency dependent force on the atom

$$\langle \vec{F}(\omega, \vec{v}) \rangle = \hbar \vec{k} \gamma \frac{I/2I_s}{1 + I/I_s + (2(\omega - \omega_0 - \vec{k} \cdot \vec{v})/\gamma)^2} \tag{6.21}$$

For a given laser intensity, the force has its maximum at $(\omega - \omega_0 - \vec{k} \cdot \vec{v}) = 0$. If the laser frequency ω is slightly red-shifted with respect to the resonance frequency ω_0 (i. e., $\omega < \omega_0$), the maximum force occurs if the particle's velocity $\vec{v}$ points in the opposite direction of $\vec{k}$ (and $\vec{F}$). In this case, the spontaneous force is more efficient in slowing the atoms down than speeding them up. Using two counterpropagating laser beams, the atoms can effectively be cooled along the common axis of the two laser beams. It is straightforward, to extend this method to three dimensions with three pairs of opposing laser beams. This situation is referred to as "Doppler-cooling" or "optical molasses" which allows to cool atomic samples down to temperatures well below 1 mK.

While the optical molasses is well suited to cool atomic samples, it does generally not allow to spatially confine them and the particles will slowly drift out of the cooling region. This fundamental constraint is known as the "optical Earnshaw theorem" formulated first by A. Ashkin and J. P. Gordon [28]. The reasoning is based on the assumption of point-like atoms whose structure (i. e., their energy levels) are fixed and independent of any other parameters. In this approximation, the force on the atoms exerted by the laser field is proportional to the Poynting vector $\vec{S}$. According to Poynting's theorem, the divergence of the Poynting vector field is

$$\vec{\nabla} \cdot \vec{S} = \frac{1}{\mu_0} \vec{\nabla} \cdot (\vec{E} \times \vec{B}) = -\vec{J}\vec{E} - \frac{1}{2} \frac{\partial}{\partial t} \left(\varepsilon_0 \vec{E}^2 + \frac{1}{\mu_0} \vec{B}^2 \right) \tag{6.22}$$

and it vanishes in the source-free trap region (i. e., for $\vec{J} = 0$) for a continuous laser field. Therefore, the force field is divergence-free, too, which means that a volume with the force pointing inwards over its entire surface cannot exist. Consequently, any trap based on the above discussed force must have a leak.

There are several loopholes to the optical Earnshaw theorem. One of them is adding a spatial dependence to the structure of the particles lifting the mere proportionality of the force to the Poynting vector. In a magneto-optical trap, this is done by introducing an inhomogeneous magnetic field which induces a position-dependent Zeeman shift to the atomic levels. For a qualitative description, we now consider a slightly more elaborate two-level atomic system with the ground state being an s-state (i. e., the orbital angular momentum is $l = 0$) and the excited state being a p-state (with $l = 1$) featuring three degenerate sublevels with the magnetic quantum numbers $m_l = -1, 0, 1$. The electron spin as well as the nuclear spin are neglected for now. This atom is placed in a magnetic field that changes linearly along the z-axis with $B_z(z) = B_z' z$. Due to the Zeeman effect, the degeneracy of the excited sublevels is lifted in the magnetic field and they split by the frequency

$$\delta(z) = \mu_B g m_l B_z(z)/\hbar \qquad (6.23)$$

with μ_B being the Bohr Magneton and g the g-factor of the excited state.

In order to trap the atoms in a small volume, two features of the system are exploited: First, the excitation rate to an individual magnetic sublevel depends on the position of the atom, because the Zeeman shift $\delta(z)$ needs to be taken into account in equation (6.21). Second, it is possible to excite selectively a transition to only one magnetic sublevel by choosing the appropriate laser polarization. Specifically, illuminating the atom with a σ^+ circularly polarized laser beam pointing in positive z-direction will only drive transitions to the $m_l = +1$ state, while a counterpropagating beam with σ^- circular polarization will drive only transitions to the $m_l = -1$ state. As depicted in Figure 6.5, the former laser beam will be in resonance for negative z and the latter one for positive z. Therefore, the choice of these laser field polarizations results in a net restoring force for a slightly red-detuned laser, i. e., the atom will always be pushed back toward the center of the trap where the magnetic field is zero.

Again, this method can be extended to three dimensions by using three pairs of opposing laser beams with appropriate polarizations. Commonly, the magnetic field gradient along each axis is generated with a pair of anti-Helmholtz coils which produce a quadrupole magnetic field that is zero at the center.

As mentioned above, the model introduced here is helpful to describe qualitatively the trapping mechanism in a magneto-optical trap, but it neglects several properties of real-world systems that can influence the trapping time, the effective potential depth, and the temperature of the trap. First, atoms have a much more complicated structure than considered above, and in particular the hyperfine-splitting results in

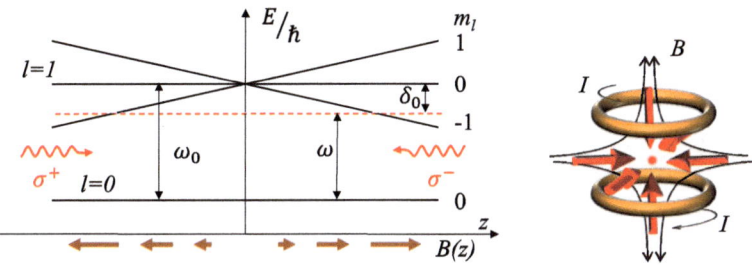

Figure 6.5: Simplified atomic energy level scheme in one dimension (left, see text) for a magnetic field created with anti-Helmholtz coils (right).

the presence of many levels close to the cooling transition. Second, the light field cannot be treated as six isolated monochromatic beams, but the intersecting beams will form a three-dimensional interference pattern which affects the local intensity and polarization on the length scale of the wavelength. And third, the coupling of different atomic levels due to the light field will induce AC Stark shifts which affect the scattering rate and result in forces along the intensity gradient of the field. Specifically in alkali atoms, which are the most common choice in laser-cooling experiments, the ground state splits into hyperfine doublets. With a typically small, but nonnegligible, branching ratio, an excited atom can decay into a hyperfine state which is not resonant to the laser radiation. Therefore, a second laser frequency is required to "repump" such atoms back into the cooling cycle. The presence of this repumper frequency makes the accurate description much more complicated because the atomic levels' dressing due to the interfering fields results in bi-chromatic forces that can significantly influence the trap dynamics (e. g., [29]).

3.2 MOTRIMS and MOTReMi

At first glance, combining magneto-optical traps with COLTRIMS is a very appealing idea. The benefits are evident: First, the very low temperature of the atoms in a MOT results in a superior momentum resolution for the recoil ions as compared to experiments with conventional gas jet targets. Second, MOTs extend the range of atomic species available for collision experiments. Low-temperature atomic beams prepared in supersonic gas jets are limited to noble gases. Commonly, MOTs work with alkali and earth-alkaline atoms which are electronically very different to noble gases having only one or two valence electrons, respectively. Third, atoms in a MOT have optically active electrons allowing to prepare a fraction of the target atoms in an excited state. This way, the initial-state dependence of the collision dynamics can be studied.

At second glance, combining MOT and COLTRIMS is not straightforward. The main obstacle is that the required magnetic fields are very different: A MOT demands a quadrupole field for the confinement of the atoms. COLTRIMS, in contrast, works best with a homogenous or vanishing magnetic field. The quadrupole field of the MOT can affect the achievable resolution of the particles' momenta or, in case of the electrons, makes a reconstruction of the momentum vector entirely impossible. In spite of this drawback, the first experiment with a magneto-optically trapped target was reported only few years after the development of COLTRIMS [30] measuring the time-of-flight of the ions. The first measurements to obtain three-dimensional recoil ion momentum vectors from a MOT target were reported by three groups at about the same time [31, 32, 33]. In these MOTRIMS experiments, the electrons were not detected, however, kinematically complete experiments were still possible for reactions with only two separated particles in the final state, e. g., electron transfer in ion-atom collisions [31, 32, 33] or photo-ionization [9, 34].

All MOTRIMS setups are based on a similar design (Figure 6.6, top). Like every COLTRIMS experiment, they comprise a recoil ion momentum spectrometer with an electric field generated by electrodes to extract the ions and direct them onto a position sensitive detector. The two anti-Helmholtz coils for the MOT field are located above and below the spectrometer electrodes, respectively, thereby avoiding electrical fringe fields which might affect the ion trajectories. One of the three pairs of cooling laser beams is oriented vertically and goes through the center of the MOT coils, the other two are oriented horizontally with an angle of 45° with respect to the direction of the spectrometer axis. As mentioned above, the quadrupole magnetic field affects the ion momentum resolution. In order to avoid this effect, some MOTRIMS experiments are operated with a switched magnetic field alternating between periods for trapping and for the momentum measurement with zero magnetic field [35].

Detecting electrons in coincidence with the recoil ions is a substantial challenge. The first experiment aiming for electron detection in a so-called MOTReMi (MOT Reaction Microscope) had a similar design to earlier MOTRIMS setups but it was supplemented with an electron detector and optimized for fast magnetic field switching [36]. However, electrons emitted directly form the MOT target couldn't be detected largely due to eddy currents induced by the fast change of the magnetic field flux [37]. In this apparatus, the MOT is loaded by a cold atomic lithium beam out of a Zeeman slower. While using this loading beam as a target instead of the magneto-optically trapped atoms allows to record electron momenta [9], this configuration results in a target temperature too high to achieve a sufficiently good recoil ion momentum resolution.

The first successful implementation of MOT and COLTRIMS measuring ion-electron coincidences was reported in 2012 by Fischer et al. [8]. The design of the apparatus differs significantly from earlier setups [7] (Figure 6.6, bottom). Most notably, the anti-Helmholtz coils are oriented coaxial to the spectrometer which allows to keep their radius smaller. Two of the laser beam pairs are oriented perpendicular to the spectrometer axis. In the standard MOT geometry, the third one would pass through the center of the coils perpendicular to the two other beam pairs. Because in the MOTReMi this direction is obstructed by the particle detectors, these beams

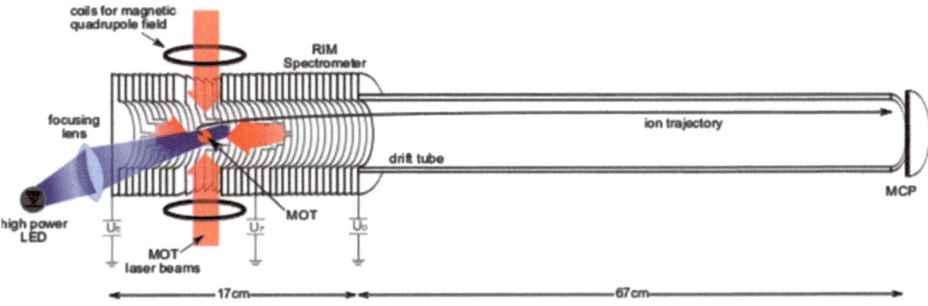

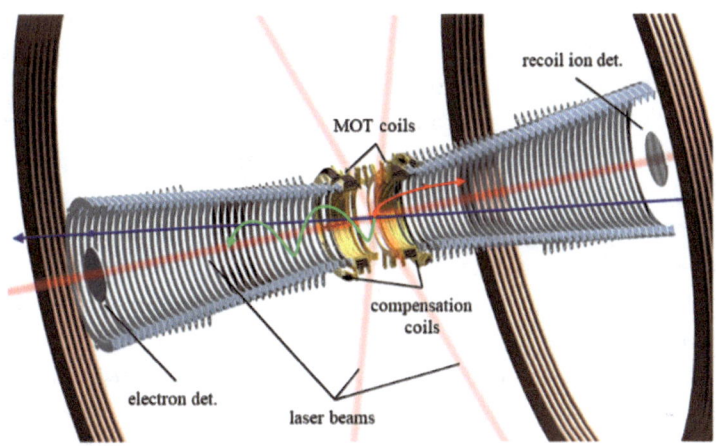

Figure 6.6: Design of a MOTRIMS setup (top, from [34]) and the MOTReMi (bottom, from [8]).

are tilted with respect to the symmetry axis by an angle of 12.5°. The anti-Helmholtz coils were supplemented with a second pair of coils slightly larger and being more distant from each other. Their purpose is to compensate the magnetic field outside of the spectrometer region close to the vacuum chamber walls, a potential source of magnetic stray fields due to eddy currents. This way, the switched magnetic field can decay to an insignificant strength in a time of about 250 µs. With the switching of the magnetic field, it becomes possible to alternate between trapping and measuring periods, each a couple of milliseconds long [7].

Recently, an alternative laser-cooling and trapping scheme was demonstrated with the same MOTReMi setup introduced above [29]. The new configuration is very similar to a standard MOT but is an all-optical trap (AOT), i. e., it does not require the inhomogeneous magnetic fields. That greatly simplifies the experimental procedure because no magnetic field switching is required. Different from the standard MOT configuration, the two opposing laser beams along each coordinate axis are geometrically not perfectly overlapped but they have a small offset to one another. The trapping mechanism is based on a complex interplay between the resonant scattering of photons as well as the position dependent shifts of the atomic energy levels "dressed" in the bi-chromatic laser field which features a complicated three-dimensional interference pattern. While this trap is not as robust as a standard MOT in the sense that it is more sensitive to small changes of the experimental parameters, it provides similar number densities and temperatures and, therefore, makes it a well-suited target for scattering experiments.

3.3 Excited and polarized atomic targets

An important feature of laser-cooled atomic targets is that they offer the opportunity to investigate ionization of ground state as well as of excited state atoms. In a magneto-optical trap, the atoms continuously undergo the transition driven by the cooling lasers resulting in an excited-state population of typically 15 % to 30 %. Therefore, ionization data taken from a laser-cooled target contain information of two possible initial states. However, it can be challenging to disentangle their cross-sections. In some experiments, the energies and momenta of all involved particles after an ionization event are measured with sufficient resolution to separate the energy levels of the initial state directly (e. g., [35, 39, 40] and Section 4.2). But this is not always possible. In particular, for very fast heavy projectiles, the energy loss of the fast particles can often only be determined with a resolution which is one or several orders of magnitude larger than the separation between the ground and excited state. In this case, the cross-sections for specific initial states have to be extracted applying a different method.

The general approach is switching the cooling lasers off periodically for a short period of time in which all the target atoms are in the ground state. During this time, the data for the ionization of the ground state I_{GS} can be acquired.[5] The time duration of the switch-off should be short as compared to the time scale of the thermal motion of the atoms in the trap, because without the laser field the damping effect of the optical molasses is ceased and the atoms move ballistically out of the trap volume and get lost for the experiment. On the other hand, the period should be much longer than the lifetime of the driven atomic transition in order to ensure that the excited state is completely depleted.

The data acquired during the time with the cooling lasers being switched on contains events for the ionization of both, the ground and excited state, delivering spectra

---

5 I_{GS} stands for any differential or summed spectrum extracted from the experimental data and it is proportional to a corresponding (differential) cross-sections.

I_{mix}. In general, the spectra for the ionization of the excited state I_{exc} can now be calculated by subtracting the two data sets:

$$I_{exc} = f \cdot I_{mix} - I_{GS} \qquad (6.24)$$

Here, the factor f depends on the excited state population P_{exc}, the ratio of the time durations T_{GS}/T_{mix} for which the data for the ground state and the mixture of the two states is obtained, as well as on fluctuations in the target atom number density during the switching cycle. If we assume that the density remains unchanged, it is $f = T_{GS}/(T_{mix} \cdot (1 - P_{exc}))$. If the total cross-sections for ground and excited state ionization σ_{GS} and σ_{exc} are known from the literature, the factor f can alternatively be calculated by $f = (1 + \sigma_{exc}/\sigma_{GS}) \cdot N_{GS}/N_{mix}$ where N_{GS} and N_{mix} are the total number of ionization events obtained in the experiment for 100 % target ground state population and for the mixture of both states, respectively. If the total cross-sections are not precisely known, it can be challenging to determine f with high accuracy, in particular if the excited state population can only be estimated. Therefore, it is important to check the spectra I_{exc} for consistency[6] and perform a careful analysis of the experimental errors including the uncertainty in f. Overall, this method works very well for those final states where the cross section of ground state ionization is smaller than the one of excited state ionization. In the opposite case, the reconstructed spectra can exhibit very large statistical errors.

In the discussion so far, we did not consider the possibility of multiple ground and excited states that can be populated in the field of the cooling lasers. In particular, different magnetic sublevels $m_l(= -l, \ldots, l)$ of the orbital angular momentum l of the active electron (i. e., the orientation of the electron's orbital angular momentum) can significantly influence the collision dynamics (see Sections 4.2, 5.1, and 6.2 in this chapter). In most MOT experiments, the trapped atomic species are alkali or alkali earth atoms with the transition between the ground s-state ($l = 0$) and an excited p-state ($l = 1$) being driven by the cooling lasers. The m_l population distribution of the excited state depends on the strength and direction of the external magnetic field, on the laser polarization and frequency, as well as on the coupling between orbital angular momentum and spins of electron and nucleus. In a conventional MOT, all possible magnetic substates are about equally populated. This is because there are no "global" magnetic eigenstates to a given quantization direction as the direction of the magnetic field vector generated by the anti-Helmholtz coils depends vastly on the exact position in the trap. Therefore, even if at a given time two separated trapped atoms would be in the same state, their temporal

---

6 A lower limit for f can be obtained considering that the derived excited spectra I_{exc} must not be negative for any region in the final state momentum space.

evolution[7] would generally be very different and depends on their position in the trap.

In general, a uniform (or vanishing) magnetic field in the trap region is required in order to obtain target polarization, i. e., the exclusive population of a specific magnetic sublevel with respect to a given quantization direction. For zero magnetic field, this can be achieved by using a single dedicated excitation laser beam. If the laser has linear polarization the p-state with $m_l = 0$ with respect to the laser polarization axis will be populated (see, e. g., [6]). Using a circularly polarized laser beam aligned to the direction of a homogenous magnetic field allows to employ optical pumping (e. g., [38, 7, 29]). In this case, even specific electronic and nuclear spin levels are populated resulting in a very high degree of polarization.

4 Electron capture in ion-atom collisions

Electron capture has extensively been studied over decades (see, e. g., [41]) due to its relevance for a large variety of research fields ranging from atomic physics, plasma physics, biophysics, and radiation damage, as well as astrophysics where it is observed, for instance, in interactions of solar wind ions with atoms. In electron capture, an electron is transferred from a bound target state to a bound state in the projectile ion. These charge exchange processes are the dominant inelastic reaction channels for slow ion-atom collisions, i. e., if the average speed of the bound electrons are much higher than the projectile speed. For sufficiently low projectile speeds, the electron cloud adjusts adiabatically to the slowly changing potential of the two nuclei.

Classically, electron capture can be understood in an "over-the-barrier" model: The electric potential along the line between the two nuclei lowers as they particles approach forming a saddle point whose height drops with decreasing distance (Figure 6.7, left). The outer electrons initially bound in the target can now move freely between the two atomic systems. As the particles veer away from each other, the height of the saddle point increases again and one or more electrons, which were initially bound in the target atom, can be captured into a bound projectile state. In quantum-mechanical descriptions, the energy potential curves of the transient molecule, which is formed during the collisions, have to be calculated. Asymptotically for large internuclear distances, the system consists initially of a neutral target atom A and the q-fold charged projectile ion B^{q+} (Figure 6.7, right). In the final state, there are two separated ions which may be in an excited state (e. g., A^+ and $B^{(q-1)+*}$). At specific distances, the

---

7 The temporal evolution of an excited state, which is generally not an eigenstate but a superposition of several nondegenerate Zeeman-shifted magnetic sublevels, can also be interpreted as the precession of the atomic magnetic moment around the magnetic field axis.

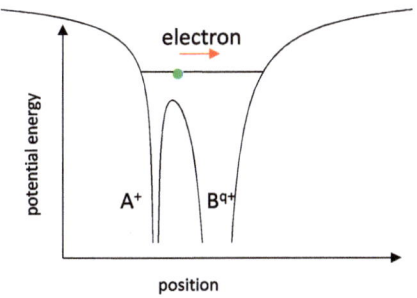 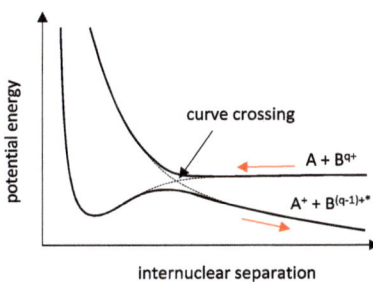

Figure 6.7: (Left) Two-center potential and electron transfer in the classical over-the-barrier model. (Right) Conceptual energy potential curves of a transient molecular ion that is formed in slow ion-atom collision. At the (avoided) curve crossing, the electron can be transferred from one potential curve to another.

corresponding potential curves can form avoided crossings, where the system undergoes the transition. The transition probabilities depend essentially on the internuclear distance at the avoided crossing and the energy gap between the potential curves.[8] They can be calculated using for instance a nonperturbative close-coupling approach (see, e. g., [44]).

In the first decade after the successful implementation of laser-cooling and momentum imaging, most MOTRIMS experiments concentrated on the measurement of electron capture. Due to energy and momentum conservation, an experiment on electron capture is already kinematically complete if the momentum vector of only a single particle, for instance the recoil ion, is measured. With its excellent recoil ion momentum resolution, MOTRIMS is ideally suited to investigate such reactions and gain detailed information on the collision dynamics.

In this section, a brief introduction into some of the concepts and ideas of electron capture measurements with MOTRIMS is given. It is out of the scope of this chapter to give a complete review of this field. For an in-depth discussion, the reader is referred to [5] which gives an excellent overview of the research done in this field until 2008. In the present context, just a few selected examples will be discussed illustrating some of the possibilities of MOTRIMS and insights that it provides into collision physics and beyond.

4.1 Collision kinematics

With respect to the kinematics, electron capture is a two-body collision with two particles in the initial (projectile ion and target atom) and two in the final state (charge-

---

[8] The model discussed here is a vast simplification and there are other transfer mechanisms that are not covered here. An excellent and concise introduction into the field can be found in [42] and [43].

changed projectile and recoiling target ion). Similar to Rutherford Scattering (cf. Section 2.1), the final momentum space—even though six-dimensional (3 projectile and 3 recoil ion momentum components)—reduces greatly if energy and momentum conservation as well as the symmetry of the reaction are considered. In the case of the Rutherford experiment, the conservation of each momentum component reduces the final state degrees-of-freedom from 6 to 3. Due to energy conservation, the number of independent final-state parameters shrinks further by one. Finally, due to cylindrical symmetry with respect to the projectile beam axis (for unpolarized initial states), the azimuthal angle of the scattered projectile, which is one of the two remaining degrees-of-freedom, does not contain any physical information and does not need to be considered. This makes the Rutherford cross section sensitive to only a single independent parameter which is commonly chosen to be the projectile scattering (solid) angle.

In spite of the similarities between electron capture and Rutherford scattering, there are, of course, significant differences. The essential distinction with respect to kinematics lies in the energy balance: Rutherford scattering corresponds to elastic collisions, while electron capture reactions are generally inelastic meaning that the total kinetic energy can differ in the initial and final state. For capture, the target atom as well as the projectile ion change their states and, thereby, their binding energies due to the transfer of the electron. The difference in total binding energies for the initial and final state is quantified by the Q-value, which is defined as $Q = E_{\text{bind},i} - E_{\text{bind},f}$.[9] Q is also referred to as "energy defect" or "inelasticity" of the collision. Generally, Q can have different discrete values depending on the target and projectile states that are populated before and after the collision.

Due to energy conservation, the inelasticity has to be balanced by the kinetic energies in the final state, and equation (6.2) can be rewritten as

$$\Delta K_{\text{proj}} = Q - \frac{1}{2} m_e v_{\text{proj},f}^2. \tag{6.25}$$

Here, ΔK_{proj} is the change in the projectile's kinetic energy excluding the kinetic energy of the captured electron which is, instead, considered in the last term on the right-hand side. It is straightforward to calculate alternatively the above kinetic energy change from the projectile momentum change:

$$\Delta K_{\text{proj}} = \frac{1}{2m_{\text{proj}}} (\vec{p}_{\text{proj},f}^2 - \vec{p}_{\text{proj},i}^2) = \frac{1}{2m_{\text{proj}}} (\vec{p}_{\text{proj},f} - \vec{p}_{\text{proj},i})(\vec{p}_{\text{proj},f} + \vec{p}_{\text{proj},i}) \approx \Delta\vec{p}_{\text{proj}} \cdot \vec{v}_{\text{proj}}. \tag{6.26}$$

---

9 The electronic binding energies have negative values in the notation of equation (6.2). The choice of the sign of Q is not consistent throughout atomic physics literature (see, e. g., [45] and [5]). In the discussion here, we use the definition given in [5] (which is widely used in nuclear physics, too), where negative Q-values correspond to a "transfer" of kinetic energy to binding energy, i. e., to endothermic reactions, and positive Q-values to the "transfer" of binding energy to kinetic energy, i. e., to exothermic reactions.

We have approximated $\vec{v}_{\text{proj},f} + \vec{v}_{\text{proj},i} \approx 2\vec{v}_{\text{proj}}$, i. e., we assume that the relative change of the projectile's velocity is insignificant. This is a very good approximation for heavy, energetic projectiles and small projectile scattering angles. Now momentum conservation is considered (cf. equation (6.1)), which can be expressed as

$$\Delta\vec{p}_{\text{proj}} = \vec{p}_{\text{proj},f} - \vec{p}_{\text{proj},i} = -(\vec{p}_{\text{tar},f} + m_e \cdot \vec{v}_{\text{proj},f}). \tag{6.27}$$

Here again, the captured electron is treated as a separate particle with a momentum of $m_e \cdot \vec{v}_{\text{proj},f}$ in the final state which does not contribute to the projectile's final momentum. Nevertheless, it should be noted that the final momenta of captured electron and projectile are linked together, because the two particles are bound and travel with identical velocity.

Combining now equation (6.25)–(6.27) results in the relation between the longitudinal momentum transfer $q_{\|}$ (which is identical to the z-component of the recoil ion momentum $p_{\text{tar},fz}$) and the Q-value:

$$Q = -q_{\|}v_{\text{proj}} - \frac{1}{2}m_e v_{\text{proj}}^2. \tag{6.28}$$

The transverse momentum balance for electron capture reactions features essentially the same kinematic behavior as Rutherford scattering. In the approximation used above, it is entirely independent of the Q-value of the reaction. It relates directly to the projectile scattering angle which is (for small scattering angles)

$$\theta_{\text{proj}} = \frac{q_{\perp}}{m_{\text{proj}}v_{\text{proj}}}. \tag{6.29}$$

A set of two independent variables used in equations (6.28) and (6.29)—for instance $(Q, \theta_{\text{proj}})$ or $(q_{\|}, q_{\perp})$—fully determine all kinematical parameters of the final state.

4.2 Initial and final state-selective cross-sections and population dynamics in a MOT

MOTRIMS enables to measure the energy balance in charge-transfer collisions in great detail. The change in the electronic binding energies, the Q-value of the reaction, can be obtained from the longitudinal recoil ion momentum by using equation (6.28). A typical longitudinal momentum spectrum (equivalent to a Q-value spectrum) is shown in Figure 6.8 for collisions of Na$^+$ ions with Rb atoms (from [35]). 6 lines can be identified, 3 of them corresponding to the capture of the rubidium 5s valence electron to the 3s, 3p, and 3d states in the sodium atom. Different initial states can be resolved, too, related to the capture from the laser-excited Rb 5p level to the Na 3p, 4s, and 4d states (denoted with an asterisk in the graph). While some of the lines overlap due to the finite experimental resolution, their relative contributions can easily be determined

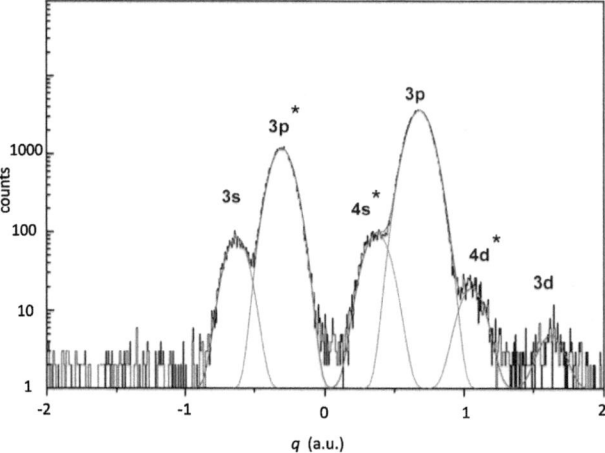

Figure 6.8: Longitudinal momentum spectrum for 2 keV Na$^+$ − Rb collision from [35] (see text).

by performing Gaussian fits on the lines. The relative cross-sections obtained in such measurements provide a sensitive test for theoretical models. In particular, the study of the state-selective cross-sections in dependence on the collision energy represents a benchmark test for classical [46, 47, 48, 49] and quantum-mechanical approximations [46, 50].

The possibility to distinguish different initial states in Q-value spectra does not only allow to investigate the collision processes themselves, but it also can be used to study the population dynamics in a magneto-optical trap. This can be achieved by carefully analyzing the lines for specific reaction channels for electron transfer from the ground state and from excited states: By switching the cooling lasers off for a given duration, a target ground state population of 100 % is obtained (see Section 3.3). When the laser is switched on again, the ground state population drops instantly and a fraction of atoms is promoted to the excited state. This drop is proportional to the relative decrease in the yield of capture reactions from the ground state atoms which can directly be observed in the Q-value spectrum. This allows to derive directly and model-independently the relative excited state population in the MOT which otherwise can only be estimated using vastly simplified models [51, 52].

A very elegant application of this technique is the time-resolved study of stimulated Raman adiabatic passage (STIRAP) reported in [53]. In STIRAP, population is coherently transferred between two states of a three-level system using two light pulses of different frequency [54]. In the specific situation studied in [53], the three levels are the 5s ground state of rubidium, the excited 5p state, and the higher excited 4d state (see Figure 6.9, left bottom). The two laser pulses are the pump pulse which is near-resonant to the 5s–5p transition, and the Stokes pulse being close to resonance to the 5p–4d excitation frequency. Both pulses are of about 50 to 70 ns duration and are delayed to one another by the time τ. Counterintuitively, population can coherently be

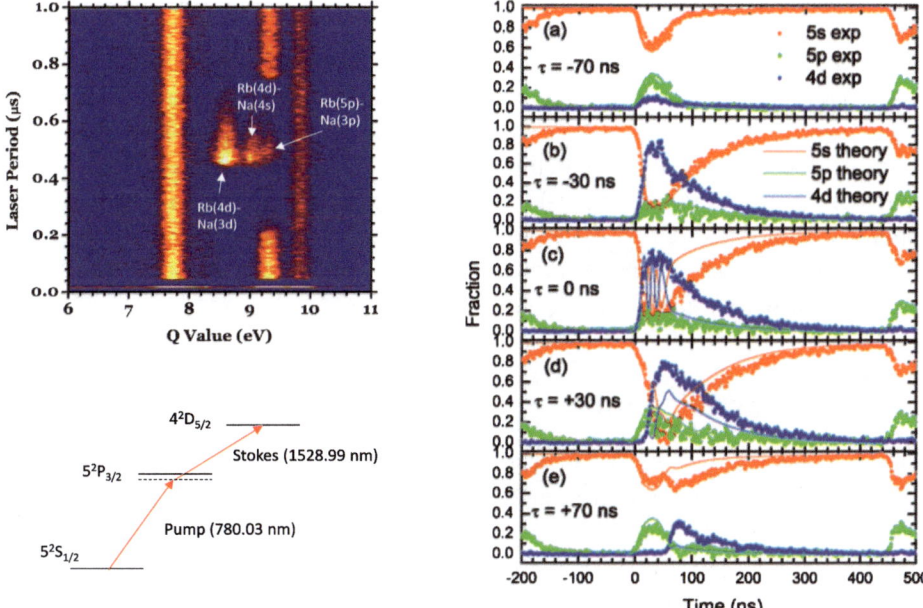

Figure 6.9: (Left, bottom) Three-level system studied in the STIRAP experiment reported in [53]. (Left, top) Electron capture yield as a function of Q-value and time in the switching cycle (from [53], see text). (Right) Relative population of ground state (red), excited 5p state (green), and higher excited 4d state (blue) (from [53]). τ is the time difference between pump and Stokes pulse. For negative τ, the Stokes pulse precedes the pump pulse.

transferred to the highest level even if the Stokes pulse slightly precedes but still overlaps with the pump pulse. Conceptually, this process can be understood in a dressed atom picture: When the Stokes pulse is switched on, the atomic levels get "dressed," i. e., there is a new set of eigenstates of the atom in the laser field. The pump pulse can then excite one of the dressed states. If the intensity of the Stokes field is adiabatically reduced, the atomic states get "undressed" again and the electron can remain in the highest excited level. Theoretically, a transfer of 100 % of the population is possible if only three atomic levels are considered and spontaneous decay is ignored. In a real experiment, very high population transfer efficiencies can be achieved, too, if the Rabi-frequencies (i. e., the laser intensities) correspond to time-scales much shorter than the variation times of the laser intensity which, in turn, should be shorter than the lifetimes of the excited states.

In the experiment reported in [53], the relative excited state populations were probed by measuring the Q-value spectrum of electron capture in 7 keV Na^+ - Rb collisions as a function of the time in the laser switching cycle (see Figure 6.9, left, top). As the cycle starts, the cooling lasers are switched on and three lines can be identified representing the capture channels Rb(5s)-Na(3p) at a Q-value of about 7.8 eV,

Rb(5p)-Na(3p) at 9.3 eV, and Rb(5s)-Na(3s) at 9.8 eV. After 200 ns, the cooling lasers are switched off and the excited state is depleted resulting in the vanishing of the line corresponding to the capture from the excited state. After about 450 ns, the Stokes and the pump pulse are switched on for a short time. Apart from the Rb(5p)-Na(3p) line that reappears, there are two additional lines at Q-values between 8 and 9 eV corresponding to the capture from the 4d level. The relative populations can now be extracted from the Q-value spectrum and are plotted as a function of time in Figure 6.9 (right). As it can be seen from the figure, the most efficient population transfer is achieved if the Stokes pulse precedes the pump pulse by 30 ns.

4.3 In angular differential cross-sections

Cross-sections differential in the Q-value and in the projectile scattering angle θ_{proj} (i. e., fully differential cross-sections FDCS), or equivalently, in the transverse momentum transfer $q_\perp$ (cf. equation (6.8)), represent the most sensitive test of collision dynamics. In a classical picture, the projectile scattering angle is closely related to the impact parameter of the collision which generally allows inferring information on the distance between the nuclei at which the charge exchange occurs. In the quantum-mechanical description, there is no one-to-one relation between scattering angle and impact parameter because the latter is principally not observable and different projectile paths contribute to the same final state. However, most calculations, which treat the electron quantum-mechanically, resort to a classical description of the projectile trajectory, too, and the transition probabilities are calculated as a function of the impact parameter. The distribution of projectile scattering angles is then obtained by a Fourier transform.[10] In this section, some recent examples are given showing how classical and quantum mechanical studies along with experimental observations allow to deepen our understanding of the dynamics in charge exchanging collisions.

In particular for large projectile charges and relatively fast collisions, the classical trajectory Monte-Carlo (CTMC) approach is a powerful method because, in contrast to coupled-channel calculations, it is not be limited by the size of the basis set, i. e., by the number of contributing quantum states which goes easily into the hundreds for these reactions, thereby increasing the required computation power tremendously. In CTMC calculations, the initial state, consisting of two atomic cores and one active electron, is randomly picked from an ensemble emulating the initial wave function. Hamilton's equations are then solved for this three-body system calculating the final state. The differential cross-sections correspond to the statistical distribution of a large number of events which are randomly generated by this method.

---

10 Note that the Fourier transform makes the calculation fully quantum-mechanical because the contributions from all impact parameters are accounted for.

In Figure 6.10, experimental and CTMC results for 2 keV/amu N^{5+}-Na(3s) collision are shown (from [49]). For the capture to the $n = 6$ shell, the agreement between experiment and CTMC is excellent. Significant quantitative deviations occur for the capture to $n = 7$, although the observed structure is qualitatively well reproduced by the calculation. Notably, the CTMC method allows to track the electron trajectories for each individual event and each capture process can be categorized according to the number of swaps that the electron undergoes between the two atomic cores. As it can be seen from the figure, a larger number of swaps results generally in larger scattering angles. In this specific study, 1 and 3 saddle point crossings of the electrons contribute significantly to the overall shape of the differential cross section. Larger numbers of swaps are negligible for the given collision energy [49].

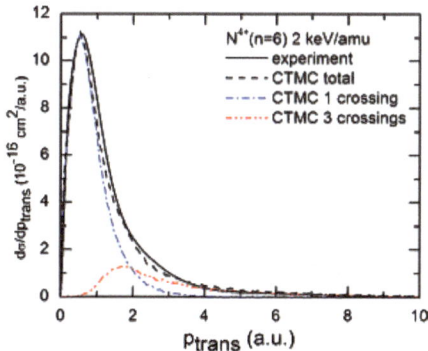 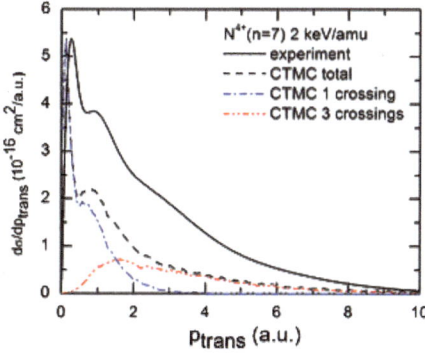

Figure 6.10: Differential cross-sections for single electron capture in 2 keV/amu N^{5+}-Na(3s) collisions (from [49]). Considered is the capture into the $n = 6$ shell (right) and the $n = 7$ shell (left). The blue (dash-dot) line corresponds the CTMC results for the electrons crossing the saddle point once, for the red line (dash-dot-dot) the electrons have three saddle-point crossings. Note, that the experimental curve (solid line) was reconstructed using a smoothing and "Abel inversion" algorithm (for details, see [49]).

The most common quantum-mechanical approaches are in some respects similar to the CTMC method discussed above: A single active electron is considered being bound in the two-center potential of target and projectile which moves along a classical trajectory. However, the electron is described quantum-mechanically as a wave function which is expressed in a carefully chosen basis set. Applying the time-dependent Schrödinger equation for a "Born-Oppenheimer" like Hamiltonian and a given impact parameter yields a set of coupled differential equations. Integration over time enables to derive the redistributed electron flux in the final state which corresponds to the state selective cross section for the chosen impact parameter. The differential cross-sections can then be obtained from the impact parameter distribution as discussed above.

In Figure 6.11, results from Leredde et al. [55] on 5 keV Na^+-Rb(5s) collisions are shown. The graph on the left depicts the redistribution of electron flux from the Rb(5s)

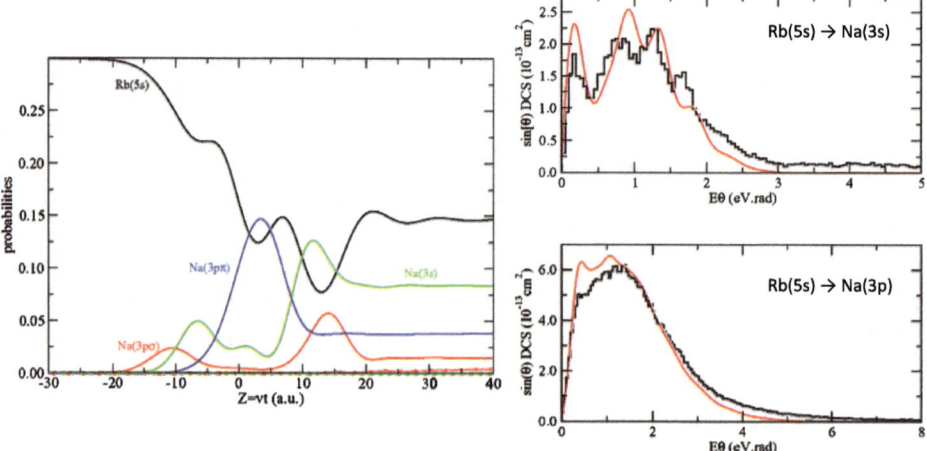

Figure 6.11: (Left) Redistribution of electron flux for 5 keV Na^+-Rb(5s) collisions and an impact parameter of 9 a. u. (from [55]). For sake of clarity, the incoming channel Rb(5s) is down-shifted by 0.7. (Right) Corresponding experimental and theoretical differential cross-sections for the capture in the Na(3s) (top) and Na(3p) (bottom) states (from [55]). The theoretical curves are from a molecular orbital close-coupling (MOCC) calculation and convoluted with the experimental resolution.

target state to the Na(3s) and Na(3p) projectile states in the course of the collision for an impact parameter of 9 atomic units and calculated in the molecular orbital close-coupling (MOCC) approximation. The differential cross-sections for these channels are shown in the graphs on the right featuring very good agreement between a quantum-mechanical theory and experiment. In particular, for the capture channel involving the final Na(3s) state, strong oscillations are observed that are already present in the impact parameter distributions (not shown in the figure). These oscillations are related to the different phases accumulated along the projectile paths for the coupled equations (or "channels") resulting in interference effects (for only two channels, these are known as "Stueckelberg oscillations"). The classical analogue of these interferences is the swaps of the electron between the nuclei discussed above.

It has been known since the 1990s that charge exchange reactions are sensitive to the orientation of the involved states. In earlier experiments, the polarization of the light emitted after charge exchange reactions was measured and a strong propensity for specific orientations of the final state were observed [56, 57]. MOTRIMS experiments enable now to orient the initial state, too, by exciting the target atoms to a specific magnetic sublevel. The propensity for orientation allows to disentangle coupling mechanisms between states of different symmetry (e. g., [58, 38]).

In a follow-up experiment to [55], the capture from the laser-excited and polarized Rb(5p) state was studied [38]. In this experiment, the excited state was prepared by using a single laser beam with circular polarization propagating along an axis perpendicular to the projectile beam direction (see Figure 6.12, left). Polarizing the tar-

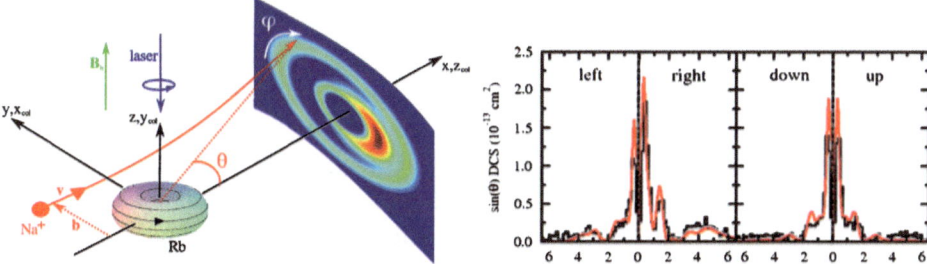

Figure 6.12: (Left) Schematics of the experiment reported in [38]. (Right) Experimental (black line) and theoretical (MOCC, red line) differential cross-sections for the reaction $Na^+ + Rb(5p, m = 1) \rightarrow Na(3p) + Rb^+$ at a projectile energy of 1 keV.

get breaks the cylindrical symmetry assumed in Section 4.1. In the differential cross-section, a considerable left-right asymmetry is observed which is well reproduced by a molecular orbital close-coupling (MOCC) calculation. In an intuitive classical picture, the enhancement of the capture probability in one direction can be explained by the velocity-matching of the electron and the projectile. The electron, which circulates classically the target nucleus, copropagates with the passing projectile on one side of the target thereby increasing the probability to be captured (Figure 6.12, right). On the opposite side, the particles are counterpropagating resulting in a reduced cross-section.

5 Energy and momentum transfer in ionizing ion-lithium collisions

Single ionization processes of atomic targets in collisions with swift ions are ideally suited to test few-body dynamics. An accurate description of the three-particle continuum can be extremely challenging despite the small number of involved particles—the projectile ion, the ejected electron, and the recoiling target ion. The simplest quantum-mechanical theoretical approach describing single ionization is the First Born Approximation (FBA) which corresponds to the first perturbative order in the projectile-target interaction. In the FBA, the target is ionized due to a single interaction between the projectile and the active electron. The perturbative expansion parameter is given by $\eta = Z_p/v_p$ with the projectile charge Z_p and its speed v_p. For small perturbation parameters $\eta \ll 1$ the FBA is expected to describe the essential aspects of the ionization process fairly well while for larger η the interaction between the nuclei and higher-order contribution in the projectile-electron scattering need to be accounted for.

Before the advent of COLTRIMS, most collision experiments which measured the electron momenta (for a review see [59]) were not capable of resolving the projectile scattering angel or the momentum transfer to the target core. Perturbative approaches such as the continuum distorted wave—eikonal initial state approximation (CDW-EIS) which account for higher-order contributions in the projectile electron interaction but in their simplest form do not consider the interaction between the heavy particles, were rather successful in describing the electron spectra even at very large perturbations ($\eta > 4$) [60] (see Figure 6.13, left). However, including the momentum transfer between the projectile and the target systems including the recoiling ion made the theoretical description significantly more complex and the approaches discussed above failed severely even if the full three body dynamics was included in the model [61] (Figure 6.13, right).

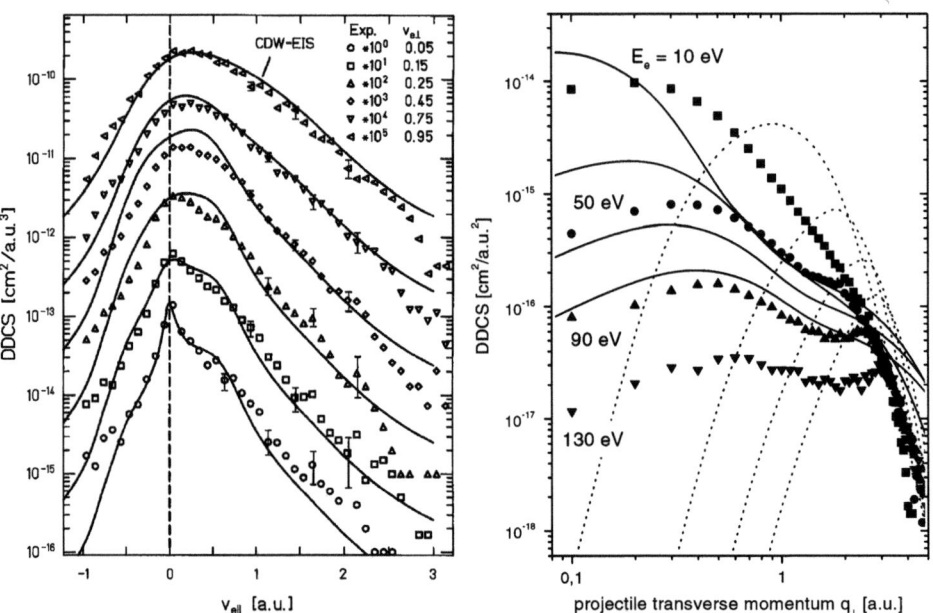

Figure 6.13: Double differential cross section for collisions of 3.6 MeV/amu Au^{53+} (i. e. η = 4.4) with atoms. (Left) The experimental cross sections are for single, double, and triple ionization of Ar and differential in the velocity components of the ejected electrons (from [60]). The solid lines are the results of the CDW-EIS calculation in a single electron picture. (Right) Cross sections differential in electron energy and projectile momentum transfer for the single ionization of He. The lines are the results of the CDW-EIS model including (solid) and excluding (dashed) the direct interaction between the nuclei.

There were several possible sources considered to be responsible for the observed disagreement: First, the approximations made to describe the projectile-target interaction could result in the observed deviations. It was shown that the details of the nucleus-nucleus interaction can have a significant impact on the shape of calculated cross-sections (e. g., [62, 63]). Second, the target is commonly described in an effective one-electron picture. Correlations between the bound electrons in the target atom are usually neglected (e. g., [64]). Third, the resolution of the measurement can affect the experimental data. It was shown that the resolution is not negligible, but it does not entirely explain the deviations [65]. Fourth, more recently the significance of the limited projectile coherence length was discovered (e. g., [66, 77]). In the theoretical treatment the projectile is described by a fully coherent wave. In the real experiments, however, the actual coherence length can be much smaller than the size of the target atoms, especially for heavy projectiles.

Earlier COLTRIMS experiments were limited to noble gas atoms or molecular targets. The development of MOTRIMS and MOTReMi made it possible to study different target species. The experiments reported below were performed with laser-cooled lithium atoms which have a very different electronic structure than noble gases. Lithium has two core electrons in the K-shell and a single valence electron in the L-shell. Electronic correlation effects can be expected to be of minor importance for the ionization of the weakly bound valence electron. The ionization of the tightly bound core electrons, in contrast, can significantly be influenced by correlations and few-electron effects. Moreover, these experiments provided two additional advancements to the earlier measurements on noble gases: First, the lower target temperature in the MOT resulted in a significantly improved momentum resolution. And second, the experiments were performed at the test ion storage ring TSR at the Max-Planck Institute for Nuclear Physics in Heidelberg, Germany. The TSR allows to achieve generally larger projectile ion coherence lengths because the ion beam is cooled by merging it with a cold electron beam. Because the coherence properties are related to the momentum spread of the beam, the cooling enables to obtain coherence lengths which are in the order of the size of the atom without losing beam intensity as it is the case for experiments at conventional single-pass beam lines. Therefore, some of the effects considered to be responsible for earlier observed discrepancies are effectively eliminated or altered in the MOTReMi measurements providing new insights.

In the following sections, double differential cross sections (DDCS) obtained in the experiments on valence ionization and inner-shell ionization for perturbation parameters of $\eta = 0.06$ (6 MeV protons) and $\eta = 1.0$ (1.5 MeV/amu O^{8+}) are discussed.

5.1 Ionization of the valence electron

In the experiments, the ionization of ground state lithium (Li(2s)) as well as of laser-excited lithium (Li(2p)) were investigated and cross-sections were extracted for fixed electron energies E_e (i. e., for a given energy transfer from the projectile to the target system) as a function of the transverse momentum transfer $q_\perp$ (equivalent to the projectile scattering angle). The experimental data represents a benchmark for theoretical models and several calculations have been employed to explain the experimental cross sections. In [67] and [68], the perturbative continuum distorted-wave eikonal-initial-state (CDW-EIS) method was applied. Nonperturbative calculations have also been performed, among them time-dependent close coupling (TDCC) [69] and coupled-pseudostate (CP) [70] approaches. Recently, also relativistic effects were considered in a nonperturbative calculation [71].

In Figure 6.14, the DDCS for the ionization of the lithium valence electron in 6 MeV proton collisions is shown. While there is a dependence of the cross sections on the initial state, there are also clear similarities between the Li(2s) and Li(2p) ionization spectra. The cross-sections feature a peak (for the lowest electron energies it is a hump) which is shifted to larger momentum transfers with increasing electron energy. The peak occurs at slightly lower momentum transfers for the 2p initial state. This peak can be interpreted as the signature of binary collisions between the projectile and the target electron and it is roughly at the position $q_\perp \approx \sqrt{2m_e(E_e + I)}$ with I being the target's ionization potential.

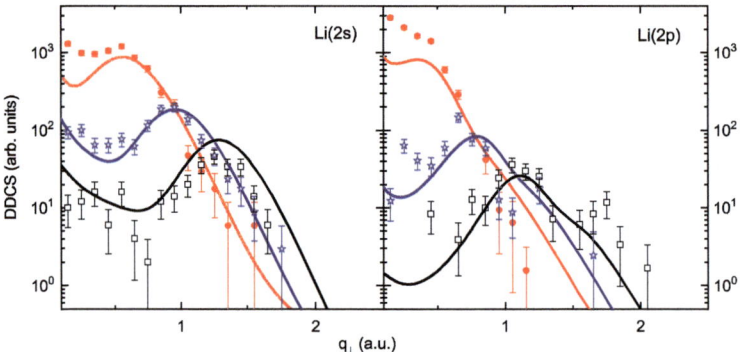

Figure 6.14: Double differential cross sections (DDCS) as a function of transverse momentum transfer for single ionization of Li(2s) (left) and Li(2p) (right) in collisions with 6 MeV protons (data from [67]). The electron energies are 2 eV (full circles), 10 eV (open stars), and 20 eV (open squares). The solid lines are the results of the CDW-EIS calculation.

For this very low perturbation ($\eta = 0.06$), all the theoretical approximation discussed above yield very similar results. Exemplarily, the results of the CDW-EIS calculation are shown in the figure. Overall, the agreement between theory and experiment is very good. There are some noticeable discrepancies at very low electron energies and at small momentum transfers which are potentially caused by both the experimental resolution and the limited projectile coherence length which was in this experiment of the order of the size of the target atom. On average, small momentum transfers correspond to larger impact parameters so that the projectile becomes increasingly incoherent relative to the effective target dimension represented by the impact parameter distribution.

For the experiment with the larger perturbation ($\eta = 1.0$) with 1.5 MeV/u O^{8+} projectiles there are distinct differences between the theoretical models (see Figure 6.15). In particular, the consideration of the nucleus-nucleus interaction has a significant effect on the shape of the cross sections. The models neglecting this interaction are systematically underestimating the cross-section for small and large momentum transfers $q_\perp$ while their yield is too large in the intermediate range between about 0.5 and 1.1 a.u. In these models, the binary mechanism discussed above prevails resulting in the local maximum at intermediate momentum transfers. The models considering the full three-body dynamics, in contrast, are in much better agreement with the experimental data in particular for the Li(2p) ionization. Some discrepancies are observed for the Li(2s) ionization at very large momentum transfers. The best result is achieved with the CP calculation which has been attributed to the more sophisticated consideration of the nucleus-nucleus interaction in this model [71].

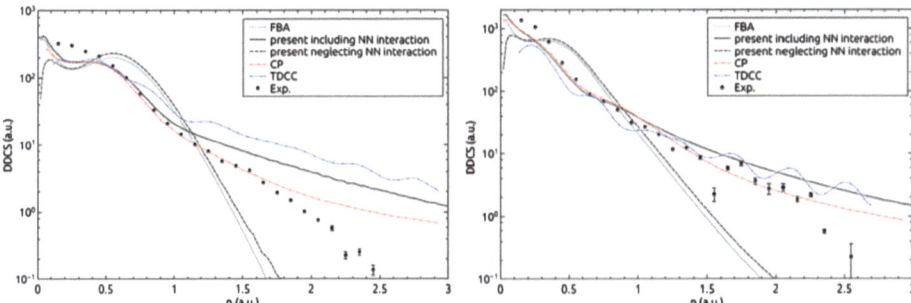

Figure 6.15: Double differential cross-sections for single ionization of Li(2s) (left) and Li(2p) (right) in collisions with 1.5 MeV/u O^{8+} projectiles (from [71]). The cross-sections are plotted as a function of the transverse momentum transfer for an electron energy of 2 eV and compared to several calculations.

Overall, the experimental data provides an excellent test for theory. For large perturbations the double-differential cross-sections are very sensitive to the full three-body dynamics. Minor deviations between the most advanced calculations and the experimental data at very small electron energies might be attributed to experimental resolution or projectile coherence effects which are not included in the models.

5.2 Core-hole creation in ionizing collisions

In collisions between projectile ions and lithium atoms, a K-shell electron (Li(1s)) can undergo a transition leaving a core-hole in the recoiling Li^+. The energy required to promote the electron from the K-shell to a higher energy level is by about 60 eV higher than the binding energy of a valence electron. Due to energy and momentum conservation, this additional energy that is transferred from the projectile to the target can easily be identified measuring the longitudinal component of the momentum transfer $q_\parallel$ from the projectile to the target system, i. e., its component in the direction of the projectile beam axis (cf. Section 4.1). In very good approximation it is $q_\parallel = (E_e + Q)/v_{proj}$ with the Q being the Q-value or 'inelasticity' of the collision, i. e., the change of binding energies of the initial and final state, and v_{proj} the projectile velocity (cf. equation (6.26)). The experimental Q-value spectrum for 1.5 MeV/u O^{8+} collisions with lithium is shown in Figure 6.16 (left). There are two well-separated peaks, one large peak with its center at about 5 eV corresponding to the ionization of the valence electron and a smaller one for the creation of a hole in the K-shell at about 65 eV. The information in the spectrum allows to distinguish unambiguously the two reaction channels.

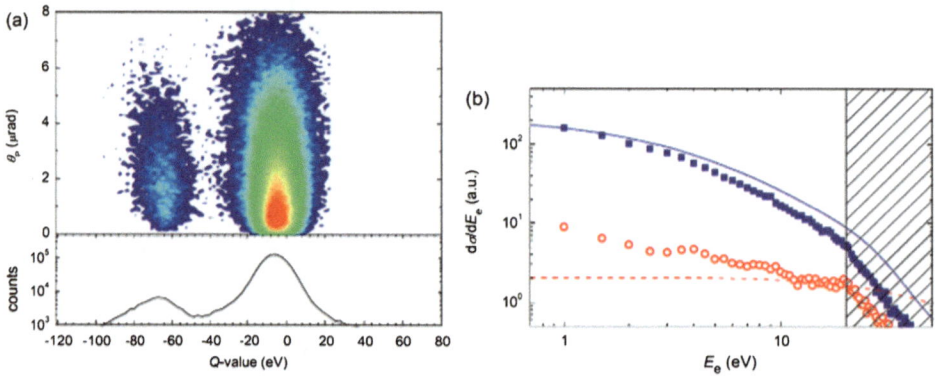

Figure 6.16: Cross-sections for the single ionization of Li(2s) in collisions with 1.5 MeV/u O^{8+} (from [8]). (Left) Cross-section as a function of Q-value and projectile scattering angle. (Right) Electron energy distribution for outer (full squares) and inner-shell (open circles) ionization compared to the CDW-EIS model (lines).

Cross-sections differential with respect to the kinetic energy of the emitted electron were extracted for the two different Q-values (Figure 6.16 (right)). The comparison with the CDW-EIS approximation reveals a surprise: The shape of the cross-sections for valence ionization is fairly well reproduced despite some deviation in the slope which is slightly steeper in the experimental data. For the core-hole creation, in contrast, there is a large margin between theory and experiment and the experimental cross-section drops significantly faster than the calculated curve over the full electron energy range.

The observed enhancement of the ionization rate for low energies was interpreted as a signature of two-electron transition [8]. The CDW-EIS model is based on a one-electron picture where the active inner-shell electron is directly promoted to the continuum. With this assumption, two important aspects are entirely neglected: First, there are different reaction channels leading to the same final state. For instance, the inner-shell electron can be excited to a bound state accompanied by the ejection of

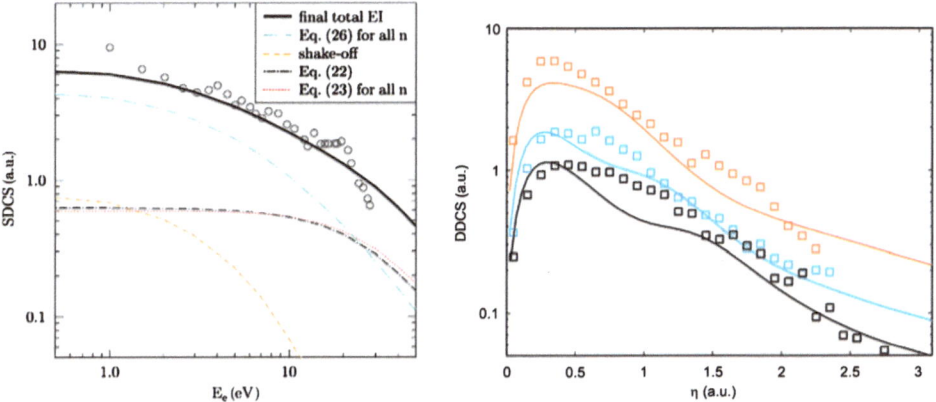

Figure 6.17: (Left) Electron energy distribution for Li inner-shell ionization in 1.5 MeV/u O^{8+} collisions (from [72]). (Right) Double differential cross-section of the same reaction for electron energies of 2 eV, 10 eV, and 20 eV (from top to bottom) (from [73]).

the weakly bound valence electron which is a process indistinguishable from the direct ionization of the Li(1s) electron. Second, the direct ionization of a K-shell electron is often accompanied by the ejection of the valence electron resulting in the double-ionization of the target atom. This reaction channel competes with the direct Li(1s) single ionization and is not included in the one-electron description of the CDW-EIS model.

Explicit few-electron processes were considered in an independent-electron model by Kirchner et al. [72]. In this approach, the projectile is described as a classical particle traveling along a straight trajectory. The dynamics of all electrons is calculated using the two-center basis-generator method (TC-BGM) combined with the CDW—EIS approach neglecting the direct correlation between the electrons. In this description, the projectile interacts with each of the three electrons (two 1s electrons and one 2s electron) individually inducing a transition or scattering elastically. In Figure 6.17 (left), the single differential cross-sections for direct Li(1s) ionization as well as for several excitation-ionization reaction channels are shown. The sum of all mechanism exhibits excellent agreement with the experimental data. Notably, shake-off—a direct correlation process between the electrons—has to be considered, too. In this process, one of the core electrons is promoted to an excited state. This transition results in the sudden change of the effective atomic potential which the initially bound valence electron is exposed to, which corresponds to a sudden change of the valence electron's eigenstates. The initial state wave function of the valence electron can now have a non-vanishing overlap with the final continuum states resulting in the ionization of that electron without requiring the direct interaction with the projectile ion. The theoretical approach was also employed to calculate double differential cross-sections [73] and the consideration of few-electron effects resulted in a significant improvement of the agreement between theory and experiment (Figure 6.17, right).

In conclusion, the process of core-hole creation in single ionization of lithium represents an ideal test bench for few-electron models. Even though the final state can be reached with only a single electron undergoing a transition, the comparison with theoretical models shows the significance of few-electron transitions and direct electron-electron correlation. In comparison to noble gas atoms, lithium is particularly well suited as a target: For helium, which has initially all its two electrons in the K-shell, there are no ambiguities between one- or two-electron transitions making it less challenging to discriminate between different reaction channels. Larger noble gas atoms such as neon or argon, on the other hand, have a much larger number of electrons which makes an explicit consideration of all possible states and transitions extremely involved. Lithium with three electrons is a system which is simple enough to include explicitly all electrons and, at the same time, it provides the possibility of complex few-electron phenomena that can be resolved in the comparison between theory and experiment.

6 Fully differential cross-sections of single ionization in ion-lithium collisions

Fully differential cross-sections (FDCS) provide the most detailed test of theoretical models (see also Chapter 2). Commonly, the FDCS are chosen to be differential in the angular distribution Ω_e of the emitted electron, in the electron energy E_e, and in the momentum transfer from the projectile to the whole target system q (or equivalently in the projectile solid angle). The remaining kinematic parameters, i. e., the momentum components of the recoil ion, are constrained by energy and momentum conservation. Notably, the first experiment succeeding in the measurement of FDCS for ion-impact induced single-ionization [24] became only possible with the development of COLTRIMS and it was reported 3 decades after similar data has been obtained for electron-impact ionization [18, 74]. The target species chosen in this first experiment was helium because it is easily cooled to sufficiently low temperatures in supersonic gas jets. It took another 9 years before FDCS for a target other than helium was measured which was lithium cooled and ionized in a MOTReMi setup [75].

For small perturbation parameters $\eta = Z_p/v_p$, i. e., for large projectile speeds v_p and not too large charges Z_p, the FDCSs for ion-impact ionization are expected to be similar to the ones for energetic electron impact ionization. In the first-order perturbative description—the First Born Approximation (FBA)—the cross-sections scale with η^2 and, therefore, become identical for protons and electrons with the same velocity. In the FBA, the angular distribution of electrons emitted from a helium target is expected to feature a two-lobe structure (see Figure 6.18, left) with most electrons emitted in the

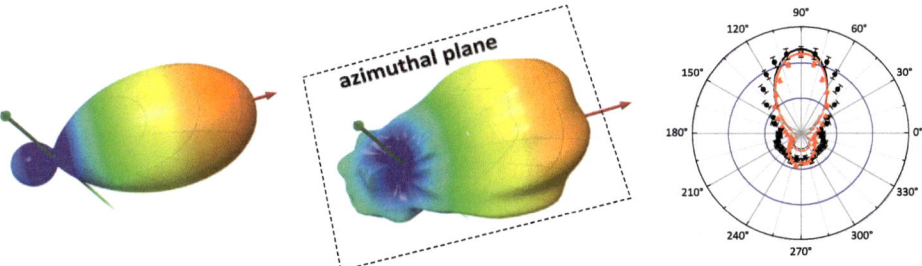

Figure 6.18: Theoretical 3DW (left) and experimental (center) fully differential cross-sections of single ionization in 1.2. GeV C^{6+} + He collisions ($\eta = 0.1$) (data from [24]). Shown is the electron emission angle in a 3D polar plot for an electron energy of 6.5 eV and a momentum transfer of 0.75 a. u. The green arrow denotes the projectile beam direction, the red the direction of the momentum transfer vector. (Right) Cut of the cross section in the azimuthal plane (black) compared to 3 MeV p + He collisions (data from [76]). Both collision systems have nearly the same perturbation parameter, but vastly different projectile coherence lengths ($\sim 10^{-3}$ a. u. and 3 a. u., respectively).

direction of the momentum transfer vector and fewer in the opposite direction. Classically, the larger peak in the direction of the momentum transfer is explained by a binary collision between the projectile and the target electron and, therefore, is often called "binary peak." Correspondingly, the smaller peak is attributed to a rescattering of the electron off the target core after colliding with the projectile, and hence, referred to as "recoil peak." The notch of the cross-section with a minimum of electron emission in the direction perpendicular to the momentum transfer is a signature of dipole-transitions which dominate the interaction between the projectile and the target.

It was a big surprise when the first experimental data became available, because despite some similarities between theoretical and experimental cross-sections the angular distribution observed experimentally was much less notched than predicted theoretically. There was much speculation about this puzzling discrepancy and, initially, it was attributed to experimental resolution [77] or the correlation between the electrons [64] in the helium target which is not accounted for in most theoretical descriptions. The nucleus-nucleus interaction, which is in principal included for instance in the 3DW approximation, was considered, too [78]. Finally, it was pointed out that the finite coherence length of the heavy projectiles might affect the collision dynamics (for reviews see [79] and Chapter 4 of this book). The influence of the projectile coherence was first observed for a molecular hydrogen target [66]. Its effect on the dynamics of helium single ionization was supported by the comparison of several experiments with different projectiles coherence lengths. The data shown in Figure 6.18 was obtained for 1.2 GeV C^{6+} projectiles whose estimated transverse coherence length was just one thousandth of the size of the atom. In a second experiment with a 3 MeV proton beam [76], the coherence length was about three times as large as the target and the agreement between theory and experiment was significantly improved though not perfect (see Figure 6.18, right), since the resolution, which was not better than for the C^{6+} experiment, still affected the measured spectra. In a recent experiment with a much better resolution and with a proton beam of large coherence length, the agreement between theory and experiment is excellent [80]. From the theoretical side, methods have been developed to account for the limited coherence length (e. g., [81, 82]) which can reproduce the structure observed in the first experiment at least qualitatively.

Fully differential data has also been obtained in the non-perturbative regime for perturbation parameters η up to more than 4 [83]. Here, the theoretical description of the ionization of helium remains a substantial challenge and essentially all theoretical approaches severely fail to reproduce experimental angular distributions (e. g., [84]). Until today, it is an open question whether the observed discrepancies have to be attributed to the finite projectile coherence length or just the incomplete account for higher-order effects in theoretical models for extreme perturbations. The experimental resolution can also have a significant impact; however, the magnitude of the discrepancies is so large that it cannot explain the discrepancies entirely.

With the development of the MOTReMi, the first kinematically complete experiment for ionization of lithium became possible. Apart from a significantly improved recoil ion momentum resolution, the target structure of lithium differs from the earlier used helium in several respects: First, because lithium has only a single valence electron, correlation effects between the target electrons are expected to be of minor importance. Second, the ionization potential is by a factor of about 4 to 5 smaller than for helium. As a result, the total ionization cross-section is much larger than for helium and the target can be ionized even at rather large impact parameters. And third, the initial wave function of the active electron is different and features a nodal structure in the radial part for the 2s ground state and—in case of the optical excitation with the cooling lasers—an angular dependence for the excited 2p state. Experiments for perturbation parameters ranging from η = 0.2 to 1.0 were performed and fully-differential cross-sections for the ionization from the 2s state and the 2p state were obtained. With this data, several questions can be investigated: What is the role of the projectile coherence length in the ion-impact induced ionization of lithium? Are there still discrepancies between theory and experiment at large perturbations, and if yes, for what reason? And finally, is there a signature of the initial electron wave function in the angular distribution of the emitted electron?

6.1 2s ground state ionization of lithium

At first glance, one might expect that the angular distribution of electrons ejected from the 2s state in lithium is similar to the one for the helium target discussed above, because the angular part of the initial state electronic wave function is identical for the two targets. This is indeed the case for small perturbation parameters η. Here, the cross-section is dominated by first-order transitions which can be interpreted as the interchange of one (virtual) photon between the projectile and the target atom which results in the dipole-like two-lobe structure in the angular distribution. In Figure 6.19, the fully differential cross-section for the ionization of lithium in collision with 2.3 MeV/amu Li^{2+} ions ($\eta = 0.2$) is shown. As expected, the angular distribution is similar to the one for the helium target and it features a binary peak in the direction of the momentum transfer and a recoil peak in the opposite direction. The relative magnitude of the recoil peak is smaller for lithium than for helium. This can be attributed to the much weaker binding of the electron making a rescattering of the electron on the target core (in a classical picture) more unlikely.

Figure 6.19: Same as Figure 6.18 (center) but for 2.3 MeV/amu Li^{2+} + Li(2s) collisions and for an electron energy of 1 eV and a momentum transfer of 0.5 a. u.

As discussed above, the finite projectile coherence length has been found to affect the differential cross-section for the helium target mainly in the direction perpendicular to the projectile beam and the momentum transfer vector, i. e., in the direction where the cross section is expected to have a node. For lithium, the node in this direction is more pronounced than for helium (cf. Figure 6.18) indicating that decoherence effects are less significant. This might have several reasons: First, the transverse ion beam coherence length was relatively large in the experiment with the lithium target because this experiment was performed at the ion storage ring TSR which allows cooling the beam to very low temperatures. As a result, incoherence effects causing the filling of the node are suppressed. Second, the enhanced visibility of the nodal structure can partly be attributed to a better experimental momentum resolution due to the much lower target temperature as compared to the earlier experiments on helium. And third, decoherence effects should generally be less significant for weaker bound electrons as can be seen by considering the limit of zero binding energy. This situation corresponds to an effective two-body scattering where quantum-mechanical (coherent) and classical (fully decoherent) differential cross-sections are known to be identical and are given by the well-known Rutherford formula. Therefore, it is plausible that the effect of the projectile coherence length diminishes with the ionization potential.

A systematic comparison of the experimental data with the 3DW-EIS approximation was performed (see Figure 6.20) for the FDCS in the azimuthal plane [85]. This theoretical approach is a fully quantum-mechanical model that includes higher-order effects, but does not account for the limited coherence length of the projectile. Overall,

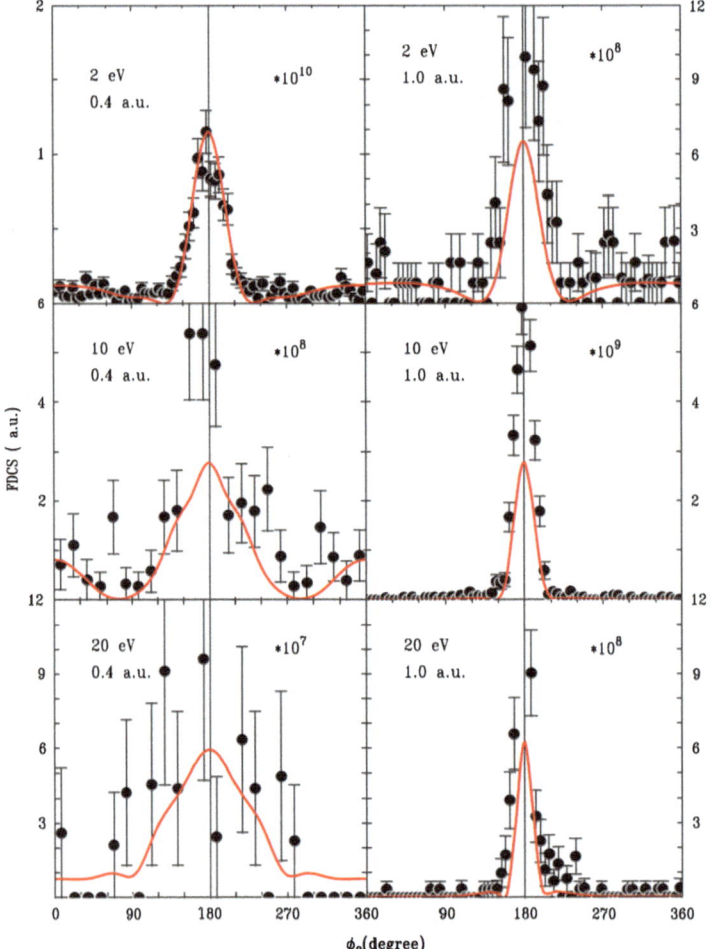

Figure 6.20: Experimental and theoretical 3DW-EIS FDCS for single ionization in 2.3 MeV/amu Li^{2+} + Li(2s) collisions as a function of the electron emission angle in the azimuthal plane (from [85]). Electron energies and momentum transfers are given in the graphs.

the shape and magnitude of the cross-sections are in good agreement with some discrepancies for large momentum transfers where the yield is underestimated by theory.

For larger perturbations, the fully-differential cross-sections can behave very different to the ones known for helium. The most distinct differences are found for large momentum transfers between the projectile and the target atom and for small kinetic energies of the ejected electron. A fully differential cross-section for ionization of lithium in 1.5 MeV/amu O^{8+} collisions is shown in Figure 6.21. Here, the projectile energy corresponds to a perturbation parameter of $\eta = 1$. The angular distribution features a large binary peak in the direction of the momentum transfer and two smaller

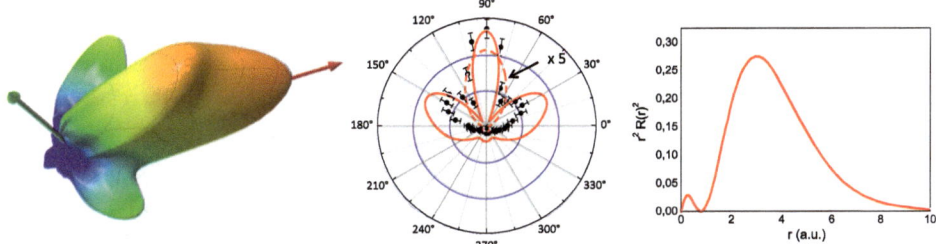

Figure 6.21: (Left) Experimental FDCS for 1.5 MeV/u O^{8+} + Li(2s) collisions for an electron energy of 1.5 eV and a momentum transfer of 1.0 a. u. (from [75]). (Center) Comparison of the FDCS with the CDW-EIS calculation in the azimuthal plane excluding (dashed line) and including (solid) the direct nucleus-nucleus interaction. (Right) Absolute square of the radial part of the Li(2s) wave function.

side-peaks shifted by about 60° with respect to the binary peak. Despite the above-discussed similarity of the 2s and 1s wave function in lithium and helium, respectively, such a three-peak structure has not been observed for a helium target. A comparison to theoretical continuum distorted wave—eikonal initial state (CDW-EIS) calculation including and excluding the direct interaction between the target nucleus and the projectile ion is shown in Figure 6.21 (center) for the azimuthal plane. Notably, the calculation reproduces the three-peak structure only if this nucleus-nucleus interaction is included.

Qualitatively, the origin of the side peaks can be understood as follows: According to the above observations there are two essential ingredients for the appearance of the side peaks; first, the 2s initial state wave function, and second, the nucleus-nucleus interaction. In contrast to the 1s state in helium, the 2s state exhibits a node in the radial part of the wave function (see Figure 6.21, right). In general, this nodal feature is of little importance because it occurs at very small distances to the lithium nucleus (< 1 a. u.) and the main contribution to the wave function is at much larger radii. However, for very close collisions, i. e., for small impact parameters, the node can become significant. For such small impact parameters, the projectile ion and the lithium nucleus are getting close and their mutual interaction intensifies resulting in a large momentum interchange between the two nuclei. Therefore, the contribution of small impact parameters to the FDCS can be enhanced by selecting large momentum transfers and small electron energies as it is done for the data in Figure 6.21. For calculated cross-sections, this effective selection of small impact parameters works only if the direct nucleus-nucleus interaction is included in the model which explains why the three-peak structure is not visible for the model excluding this interaction. Hence, the side peaks can then be interpreted as the signature of the nodal structure in the 2s wave function which is "probed" by the projectile ion for collisions with small impact parameters.

The experimental data provides a benchmark test for theoretical calculations and several models have been applied (see Figure 6.22). While all the calculations consider the nucleus-nucleus interaction as well as other higher-order contributions they differ significantly with respect to the approximations used. Hubele et al. [75] and Ghanbari-Adivi et al. [86] employ the CDW-EIS and 3DW-EIS approaches, respectively, describing higher-order effects using distorted waves or eikonal phases to approximate the particles' wave functions in the final and initial state; Walters et al. [70] and Gulyas et al. [68] use an impact parameter approach describing the projectile in coordinate space obtaining the momentum distributions by Fourier transformation. While Walters et al. employ couple pseudostates (CP) to approximate the continuum states, Gulyas et al. are using an approximated analytic polarization potential in order to account for higher-order effects in the projectile target interaction. All the approximations reproduce the experimentally observed three-peak structure as well as the widths and positions of the peaks. There are some differences in the relative height of the side-peaks which depends sensitively on the details of the calculation. However, it can be concluded that experiment and models are in reasonable agreement and, at least up to a perturbation parameter of $\eta = 1$, the developed theoretical methods reproduce successfully the ionization dynamics.

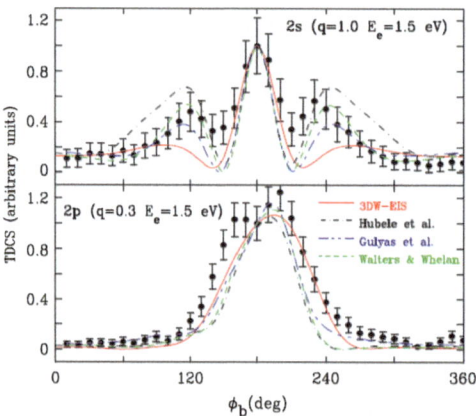

Figure 6.22: Experimental FDCS in the azimuthal plane compared to several theoretical models (see text) for the single ionization of Li(2s) (top) and Li(2p) (bottom) in collisions with 1.5 MeV/u O^{8+} (graph from [86]).

6.2 2p excited state ionization of lithium

In the optical field of the cooling lasers, typically 15 % to 35 % of all lithium atoms are promoted to the 2p state which allows to obtain FDCS for the ionization of excited atoms. In comparison to the ground state, the ionization potential of the 2p state is by about 35 % smaller resulting in an overall larger total ionization cross-section for excited atoms. In contrast to the s-states in lithium and helium, the 2p wave function is generally not spherically symmetric, because the orbital angular momentum vector $\vec{L}$ is nonzero and can have a well-defined z-component $L_z = M_L \hbar$ with respect to a specific quantization direction z with the magnetic quantum number being $M_L = -1$, 0, or +1. As discussed in Section 3.3 of this book chapter, the direction of the quantization axis and the population distribution among the different magnetic sublevels (i. e., the degree of polarization) depends on the direction of the external magnetic field and the polarization and detuning from resonant energy of the laser beams used for excitation. In the experiments discussed below, the magnetic field direction coincides with the projectile ion beam direction and the laser beam properties result in a predominant population of the $M_L = -1$ state corresponding to the wave function shown in Figure 6.23 (right).

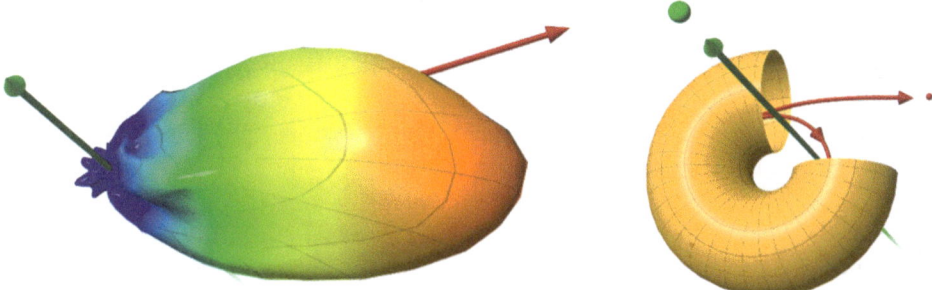

Figure 6.23: (Left) FDCS for 2.3 MeV/amu Li^{2+} + Li(2p) collisions for an electron energy of 1 eV and a momentum transfer of 0.5 a. u. (from [75]). (Right) Depiction of the electron wave function for the Li(2p) state with $M_L = -1$. The projectile momentum vector corresponds to the green arrow. A classical electron trajectory is shown in red.

The experimental fully differential cross-section for small perturbation parameters in collisions with 2.3 MeV/amu Li^{2+} ions is shown in Figure 6.23 (left). The electron angular distribution features a single binary peak indicating that the ionization dynamics is vastly dominated by the direct interaction of the projectile ion with the target electron. Most notably, the binary peak is not oriented in the direction of the momentum transfer vector as one would expect for such a pure binary collision scenario. Rather, the peak is shifted in the azimuthal plane by an angle of about 15°. Due to symmetry considerations, such a shift can generally not occur for spherically symmetric initial states (such as s-states or unpolarized p-states), because in this case there is only a single vector that breaks the overall spherical symmetry—this is the velocity vector of the incoming projectile. Therefore, all other momentum vectors of the individual particles are distributed around the projectile beam axis featuring cylindrical symmetry. Relative azimuthal angles, for instance between the electron momentum vector and the momentum transfer vector do not have a preferred handedness, i. e., φ and $-\varphi$ occur with the same probability. As a consequence, cross-sections differential in the (relative) electron emission angle have to be symmetric to the plane spanned by the projectile velocity and momentum transfer vectors (commonly referred to as "scattering plane"). The observed angular shift of the FDCS (Figure 6.23) has to be attributed to the polarization of the target. The predominant excitation of a specific magnetic sublevel results in a second distinct direction given by the orbital angular momentum of the initial state wave function $\vec{L}$ which breaks the cylindrical symmetry.

For the experiment discussed here, the initial electronic probability density (i. e., the absolute square of the wave function) has a toroidal shape as shown in Figure 6.23 (right), with its symmetry axis coinciding with the projectile beam direction. While the amplitude of the electronic wave function still features cylindrical symmetry to the projectile beam axis, the phase of the wave function (or the handedness of the angular momentum) breaks the azimuthal symmetry and results in the observed shift of the angular distribution. More intuitively, this can be understood in terms of the probability current density $\vec{j} = \hbar/(2mi)(\psi^*\nabla\psi - \psi\nabla\psi^*)$ which "flows" with a distinct direction around the origin in analogy to a "classical" electron circling around the nucleus in the Bohr model (red trajectory in Figure 6.23). In the process of the ejection of the electron, the signature of this rotational direction does not get lost but it is imprinted in the final state angular distribution.

In a simple classical picture, the direction of the angular shift can qualitatively be understood considering the motion of the electron only in the azimuthal plane. First, we assume that the electron moves initially clockwise around the target nucleus with a momentum of $\vec{p}_{el,i}$ as depicted in Figure 6.24 (bottom). In the collision, the projectile transfers the momentum $\vec{q}_\perp$ and accelerates the electron to the momentum $\vec{p}_{el,f}$ such that the electron gains sufficient kinetic energy to escape the atom (nucleus-nucleus interaction is neglected in this picture). However, due to the attractive potential of the target core, the electron is not emitted in the direction of the momentum transfer but its emission direction is shifted to the right in accordance with the experimental observation (see left figure). In the example discussed so far, the initial electron momentum and the momentum transfer vector point in the same direction. Of course, this is not always the case and any angle between the two vectors is possible. A second example is depicted in the figure on the right where $\vec{p}_{el,i}$ and $\vec{q}_\perp$ are in opposite directions. It should be noted, that here the momentum transfer has to be significantly larger in order for the electron to gain sufficient energy to escape the attractive potential of the target nucleus. In this scenario, the electron angle is shifted to left by the target potential which is again in accordance to the experimental observations.

The cross-sections for 2p-ionization were systematically compared to the results of the 3DW-EIS approximation [85] (see Figure 6.25). Overall, satisfactory agreement between theory and experiment is found, and the widths and shifts of the angular distributions are well reproduced while there are some discrepancies in the relative magnitude of the peaks. Notably, for electron energies of 10 eV and momentum transfers of 1 a. u. there is a double-peak structure predicted by theory which is not visible in the experimental data. This structure has been traced back [85] to the angular part

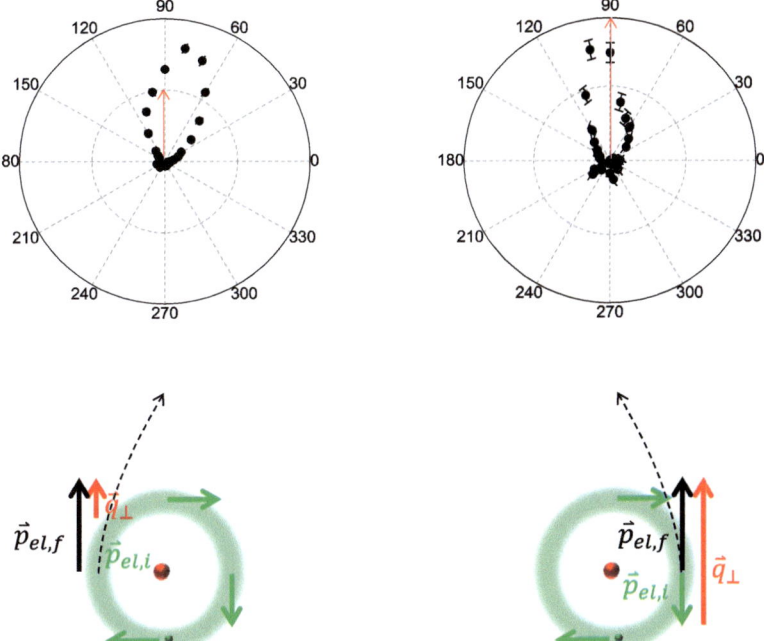

Figure 6.24: (Top) FDCS as a function of the electron ejection angle in the azimuthal plane for target single ionization in 2.3 MeV/amu Li^{2+} + Li(2p) collisions. Electron energy is 2 eV and momentum transfers are 0.4 a. u. (left) and 1.0 a. u. (right), respectively. The momentum transfer direction is denoted by the red arrow. (Bottom) Classical picture of electron ejection. The electron (green) circle clockwise around the nucleus (red). After momentum is transferred (red arrow), the electron leaves the atom along the dashed trajectory.

of the initial state wave function, but the feature is too narrow to be resolved in the experimental data.

For the larger perturbation parameter of $\eta = 1$ in 1.5 MeV/u O^{8+} collisions, a fully differential cross-section is shown in Figure 6.22 (bottom) and compared to the same theoretical models discussed in the previous section [86]. Here, too, the electron angular distribution exhibits an asymmetry to the momentum transfer direction due to the polarization of the target. The theoretical models are in good agreement with each other and reproduce the experimental spectrum with respect to peak position very well while the experimentally observed width is slightly underestimated.

In conclusion, the fully-differential cross-sections obtained for the lithium target show substantial differences to earlier obtained data for a helium target. Most importantly, the FDCSs exhibit a strong dependence on the details of the target initial state. Available theoretical approximations reproduce the experimental data with very good agreement for perturbation parameters up to $\eta = 1$. Even though projectile coherence effects are probably weaker than for earlier data with helium as a target, the limited co-

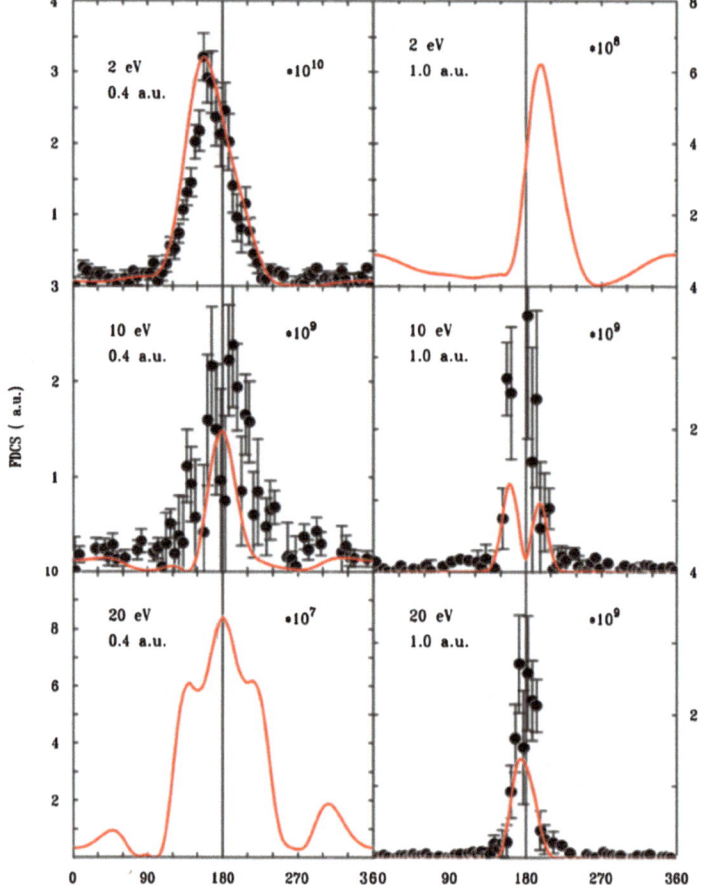

Figure 6.25: Same as Figure 6.22 but for the ionization of the Li(2p) state (from [85]).

herence length might still contribute to small discrepancies between experiment and theory. However, no conclusive statement about the importance of coherence effects can be made here.

7 Concluding remarks

In this book chapter, the concepts and methodologies of COLTRIMS, MOTs, and the combination of both in MOTRIMS and MOTReMi are discussed mainly in context with their application in ion-atom scattering experiments. It is worth noting that momentum imaging in experiments with laser-cooled targets is a much wider field than highlighted here both with respect to experimental techniques as well as regarding the physical systems under investigation. On the technical side, there are experiments using laser-cooled targets without employing magneto-optical traps. Alternatively, the target can be provided by laser-cooled atomic beams [9] or it can be trapped in an optical dipole trap [87]. In other studies (e. g., [88, 89]), laser cooling was combined with traditional electron spectrometers (see Figure 6.1) rather than with COLTRIMS, and these experiments also provide kinematically complete information on scattering processes.

Atomic collision processes under investigation include—apart from ionization and capture in ion-atom collisions, as discussed in this chapter—photoionization of atoms by synchrotron [90] and free-electron laser radiation [6], as well as few [88] and multiphoton ionization by optical laser radiation [9]. Kinematically complete experiments on electron impact ionization (e, 2e) have been performed [89], too. These studies often yield a superior resolution than experiments with conventionally cooled targets due to much lower target temperatures that are achievable by laser cooling. Moreover, the atomic target species (mainly alkali atoms) that are trapped in MOTs have a very different electronic structure than noble gases studied in most COLTRIMS experiments using supersonic gas jets. Specifically, alkali atoms have the distinct characteristics that they possess only a single valence electron for which correlations effects with other bound electrons are much weaker than, e. g., in noble gases. Because the binding energy of this valence electron is small, the electronic dynamics is substantially "slower" which, in many situations, can simplify the investigation of a specific process. These particular features can be exploited in a large variety of experiments including all types of projectiles. As an example, the study of few-photon ionization reactions, e. g., of helium require (X)UV radiation and/or attosecond pulses which are afflicted with substantial experimental complications. Corresponding studies can be done with alkali atoms just with relatively simple optical or infrared table-top lasers (for a recent example, see [91]).

MOTRIMS is not limited to atomic physics studies, but it has been applied to investigate nuclear decay reactions of radioactive isotopes (e. g., [10]). These studies principally allow for a detailed investigation of symmetries of the standard model (for a review, see [92]). Present efforts are even aiming for the discovery of sterile neutrinos [93] which are among the possible candidates to explain galactic dark matter. These particles are hoped to be traceable by measuring momenta and energies of electrons, ions, and photons in coincidence in beta-decay or K-shell capture reactions. The few examples listed here illustrate that the possibilities opened up by combining laser cooling and momentum imaging are by far not yet exhausted.

Bibliography

[1]   Metcalf HJ, van der Straten P. Laser Cooling and Trapping. New York: Springer-Verlag; 1999.
[2]   Metcalf HJ, van der Straten P. Atoms and Molecules interacting with light. Cambridge University Press; 2016.

[3]   Ullrich J, Moshammer R, Dorn A, Dörner R, Schmidt LPH, Schmidt-Böcking H. Recoil-ion and electron momentum spectroscopy: reaction-microscopes. Rep. Prog. Phys. 2003;66:1463.
[4]   Dörner R, Mergel V, Jagutzki O, Spielberger L, Ullrich J, Moshammer R, Schmidt-Böcking H. Cold Target Recoil Ion Momentum Spectroscopy: a 'momentum microscope' to view atomic collision dynamics. Phys. Rep. 2000;330:95.
[5]   DePaola BD, Morgenstern R, Andersen N. MOTRIMS: Magneto–Optical Trap Recoil Ion Momentum Spectroscopy. Adv. At. Mol. Opt. Phys. 2008;55:139.
[6]   Zhu G, Schuricke M, Steinmann J, Albrecht J, Ullrich J, Ben-Itzhak I, Zouros TJM, Colgan J, Pindzola MS, Dorn A. Phys. Rev. Lett. 2009;103:103008.
[7]   Hubele R, Schuricke M, Goullon J, Lindenblatt H, Ferreira N, Laforge A, Brühl E, Bastos de Jesus VL, Globig D, Kelkar A, Misra D, Schneider K, Schulz M, Sell M, Song Z, Wang X, Zhang S, Fischer D. Electron and recoil ion momentum imaging with a magneto-optically trapped target. Review of Scientific Instruments. 2015;86(3):033105.
[8]   Fischer D, Globig D, Goullon J, Grieser M, Hubele R, de Jesus VLB, Kelkar A, LaForge A, Lindenblatt H, Misra D, Najjari B, Schneider K, Schulz M, Sell M, Wang X. Ion-Lithium Collision Dynamics Studied with a Laser-Cooled In-Ring Target. Phys. Rev. Lett. 2012;109:113202.
[9]   Schuricke M, Zhu G, Steinmann J, Simeonidis K, Ivanov I, Kheifets A, Grum-Grzhimailo AN, Bartschat K, Dorn A, Ullrich J. Strong-field ionization of lithium. Phys. Rev. A. 2011;83:023413.
[10] Hong R, Leredde A, Bagdasarova Y, Fléchard X, García A, Knecht A, Müller P, Naviliat-Cuncic O, Pedersen J, Smith E, Sternberg M, Storm DW, Swanson HE, Wauters F, Zumwalt D. Phys. Rev. A. 2017;96:053411.
[11] Geiger H. On the Scattering of α-Particles by Matter. Proceedings of the Royal Society of London A. 1908;81:174–177.
[12] Geiger H, Marsden E. On a Diffuse Reflection of the α-Particles. Proceedings of the Royal Society of London A. 1909;82:495–500.
[13] Geiger H. The Scattering of the α-Particles by Matter. Proceedings of the Royal Society of London A. 1910;83:492–504.
[14] Bothe W, Geiger H. Z. Physik. 1924;26:44.
[15] Rutherford E, Robinson H. Phil. Mag. 1913;26:717.
[16] Robinson HR. Phil. Mag. 1925;50:241.
[17] Siegbahn Kai, Prize Nobel. Nobel Lectures 1981, Physics 1981–1990. Singapore: World Scientific Publishing Co.; 1993.
[18] Ehrhardt H, Schulz M, Tekaat T, Willmann K. Phys. Rev. Lett. 1969;22:89.
[19] Maydanyuk NV et al. Phys. Rev. Lett. 2005;94:243201.
[20] Ullrich J, Schmidt-Böcking H. Phys.Lett. A. 1987;125:193.
[21] Jagutzki O. PhD thesis, University of Frankfurt.
[22] Dörner R et al. Phys. Rev. Lett. 1994;72:3166.
[23] Moshammer R, et al. Phys. Rev. Lett. 1994;73:3371.
[24] Schulz M, et al. Nature. 2003;422:48.
[25] Moshammer R, Fischer D, Kollmus H. Recoil-Ion Momentum Spectroscopy and "Reaction Microscopes". In: Ullrich J, Shevelko VP, editors. Many-Particle Quantum Dynamics in Atomic and Molecular Fragmentation. Berlin, Heidelberg, New York: Springer-Verlag; 2003.
[26] Wiley WC, McLaren IH. Rev. Sci. Instrum. 1955;26:1150.
[27] Raab EL, Prentiss M, Cable Alex, Chu Steven, Pritchard DE. Phys. Rev. Lett. 1987;59:2631.
[28] Ashkin A, Gordon JP. Opt. Lett. 1983;8:511.
[29] Sharma S, Acharya BP, De Silva AHNC, Parris NW, Ramsey BJ, Romans KL, Dorn A, de Jesus VLB, Fischer D. Phys. Rev. A. 2018;97:043427.
[30] Wolf S, Helm H. Phys. Rev. A. 1997;56:R4385.
[31] van der Poel M, Nielsen CV, Gearba M-A, Andersen N. Phys. Rev. Lett. 2001;87:123201.

[32] Turkstra JW, Hoekstra R, Knoop S, Meyer D, Morgenstern R, Olson RE. Phys. Rev. Lett. 2001;87:123202.
[33] Flechard X, Nguyen H, Wells E, Ben-Itzhak I, DePaola BD. Phys. Rev. Lett. 2001;87:123203.
[34] Götz S, Höltkemeier B, Hofmann CS, Litsch D, DePaola BD, Weidemüller M. Rev. Sci. Instrum. 2012;83:073112.
[35] Blieck J, Flechard X, Cassimi A, Gilles H, Girard S, Hennecart D. Rev. Sci. Instrum. 2008;79:103102.
[36] Steinmann Jochen. PhD Thesis, Heidelberg.
[37] Schuricke Michael. PhD Thesis, Heidelberg.
[38] Leredde A, et al. Phys Rev Lett. 2013;111:133201.
[39] Bredy R et al. Nucl. Instr. and Meth. in Phys. Res. B. 2003;205:191.
[40] Shah MH, Camp HA, Trachy ML, Veshapidze G, Gearba MA, DePaola BD. Phys. Rev. A. 2007;75:053418.
[41] Bransden BH, McDowell MRC. Charge Exchange and the Theory of Ion-Atom Collisions. Oxford: Clarendon; 1992.
[42] Nikitin Evgueni E. Adiabatic and Diabatic Collision Processes at Low Energies. In: Drake GWF, editor. Springer Handbook of Atomic, Molecular, and Optical Physics. Berlin, Heidelberg, New York: Springer-Verlag; 2006.
[43] Gargaud Muriel, McCarroll Ronald. Ion–Atom Charge Transfer Reactions at Low Energies. In: Drake GWF, editor. Springer Handbook of Atomic, Molecular, and Optical Physics. Berlin, Heidelberg, New York: Springer-Verlag; 2006.
[44] Ovchinnikov SY, Ogurtsov GN, Macek JH, Gordeev YS. Phys. Rep. 2004;389:119.
[45] Shevelko VP, Ullrich J. Kinematics of Atomic and Molecular Fragmentation Reactions. In: Ullrich J, Shevelko VP, editors. Many-Particle Quantum Dynamics in Atomic and Molecular Fragmentation. Berlin, Heidelberg, New York: Springer-Verlag; 2003.
[46] Knoop, S. "Electron dynamics in ion–atom interactions." PhD thesis, Rijksuniversiteit Groningen. 2006.
[47] Knoop, S, Olson, RE, Ott, H, Hasan, VG, Morgenstern, R, Hoekstra, R. Single ionization and electron capture in He2+ + Na collisions. J. Phys. B: At. Mol. Opt. Phys. 2005;38:1987.
[48] Otranto S, Hoekstra R, Olson RE. Phys. Rev. A. 2014;89:022705.
[49] Blank I, Otranto S, Meinema C, Olson RE, Hoekstra R. Phys. Rev. A. 2015;87:032712.
[50] Knoop, S, Keim, M, Lüdde, HJ, Kirchner, T, Morgenstern, R, Hoekstra, R. State selective single-electron capture in O6+ + Na collisions. J. Phys. B: At. Mol. Opt. Phys. 2005;38:3163.
[51] Shah MH, Camp HA, Trachy ML, Veshapidze G, Gearba MA, DePaola BD. Physical Review A. 2007;75(5):053418.
[52] Veshapidze G, Bang J-Y, Fehrenbach CW, Nguyen H, DePaola BD. Phys. Rev. A. 2015;91:053423.
[53] Gearba MA, Camp HA, Trachy ML, Veshapidze G, Shah MH, Jang HU, DePaola BD. Phys. Rev. A. 2007;76:013406.
[54] Bergmann K, Theuer H, Shore BW. Rev. Mod. Phys. 1998;70:1003.
[55] Leredde A, Cassimi A, Fléchard X, Hennecart D, Jouin H, Pons B. Phys. Rev. A. 2012;85:032710.
[56] Roncin P, Adjouri C, Gaboriaud MN, Guillemot L, Barat M, Andersen N: Phys. Rev. Lett. 1990;65:3261.
[57] Roncin P, Adjouri C, Andersen N, Barat M, Dubois A, Gaboriaud MN, Hansen JP, Nielsen SE, Szilagyi SZ: J. Phys. B. 1994;27:3079.
[58] Gargaud M, Bacchus-Montabonel MC, Grozdanov T, McCarroll R: J. Phys. B. 1994;27:4675.
[59] Stolterfoht Nikolaus, DuBois Robert D, Rivarola Roberto D. Electron Emission in Heavy Ion-Atom Collisions. Berlin, Heidelberg, New York: Springer-Verlag; 1997.
[60] Moshammer R, Fainstein PD, Schulz M, Schmitt W, Kollmus H, Mann R, Hagmann S, Ullrich J Phys Rev. Lett. 1999;83:4721.

[61]   Moshammer R, Perumal A, Schulz M, Rodriguez V D, Kollmus H, Mann R, Hagmann S, Ullrich J
       Phys. Rev. Lett. 2001;87:223201.
[62]   Madison DH, Fischer D, Foster M, Schulz M, Moshammer R, Jones S, Ullrich J Phys. Rev. Lett.
       2003;91:253201.
[63]   Foster M, Peacher JL, Schulz M, Madison DH, Chen Z, Walters HRJ Phys. Rev. Lett.
       2006;97:093202.
[64]   Fiol J, Olson RE. J. Phys. B. 2004;37:3947.
[65]   Dürr M, Najjari B, Schulz M, Dorn A, Moshammer R, Voitkiv AB, Ullrich J. Phys. Rev. A.
       2007;75:062708.
[66]   Egodapitiya KN, Sharma S, Hasan A, Laforge AC, Madison DH, Moshammer R, Schulz M. Phys.
       Rev. Lett. 2011;106:153202.
[67]   LaForge AC, Hubele R, Goullon J, Wang X, Schneider K, de Jesus VLB, Najjari B, Voitkiv AB,
       Grieser M, Schulz M, Fischer D. J Phys B. 2013;46:031001.
[68]   Gulyás L, Egri S, Kirchner T. Phys. Rev. A. 2014;90:062710.
[69]   Ciappina MF, Pindzola MS, Colgan J Phys. Rev. A. 2013;87:042706.
[70]   Walters HRJ, Whelan Colm T. Phys. Rev. A. 2014;89:032709.
[71]   Bondarev AI, Kozhedub YS, Tupitsyn II, et al. Eur. Phys. J. D. 2019;73:46.
       doi:10.1140/epjd/e2019-90635-9.
[72]   Kirchner T, Khazai N, Gulyás L. Phys. Rev. A. 2014;89:062702.
[73]   Spiewanowski MD, Gulyás L, Horbatsch M, Kirchner T. Phys. Rev. A. 2016;93:012707.
[74]   Schulz, M et al. Triply differential single ionization cross-sections in coplanar and
       non-coplanar geometry for fast heavy ion-atom collisions. J. Phys. B. 2001;34:L305–L311.
[75]   Hubele R, LaForge A, Schulz M, Goullon J, Wang X, Najjari B, Ferreira N, Grieser M, de Jesus
       VLB, Moshammer R, Schneider K, Voitkiv AB, Fischer D. Polarization and Interference Effects in
       Ionization of Li by Ion Impact. Phys. Rev. Lett. 2013;110:133201.
[76]   Wang X, Schneider K, LaForge A, Kelkar A, Grieser M, Moshammer R, Ullrich J, Schulz M,
       Fischer D. J. Phys. B. 2012;45:211001.
[77]   Fiol J, et al. J. Phys. B. 2006;39:L28.
[78]   Schulz M, Dürr M, Najjari B, Moshammer R, Ullrich J. Phys. Rev. A. 2007;76:032712.
[79]   Schulz M. Advances in Atomic, Molecular, and Optical Physics. 2017;66:507–543.
[80]   Gassert H, Chuluunbaatar O, Waitz M, Trinter F, Kim H-K, Bauer T, Laucke A, Müller Ch,
       Voigtsberger J, Weller M, Rist J, Pitzer M, Zeller S, Jahnke T, Schmidt LPhH, Williams JB, Zaytsev
       SA, Bulychev AA, Kouzakov KA, Schmidt-Böcking H, Dörner R, Popov YuV, Schöffler MS. Phys.
       Rev. Lett. 2016;116:073201.
[81]   Járai-Szabó, F, Nagy, L. Eur. Phys. J. D. 2015;69:4.
[82]   Sarkadi L. Phys. Rev. A. 2018;97:042703.
[83]   Fischer D, et al. J. Phys. B. 2003;36:3555.
[84]   McGovern M et al. Phys. Rev. A. 2010;81:042704.
[85]   Ghanbari-Adivi E, et al. J. Phys. B. 2017;50(21):215202.
[86]   Ghanbari-Adivi Ebrahim, Fischer Daniel, Ferreira Natalia, Goullon Johannes, Hubele Renate,
       LaForge Aaron, Schulz Michael, Don Madison. Phys. Rev. A. 2016;94:022715.
[87]   Dorn A. Private communication.
[88]   Pursehouse James, Murray Andrew James, Wätzel Jonas, Berakdar Jamal. Phys. Rev. Lett.
       2019;122:053204.
[89]   Nixon Kate L, Murray Andrew James. Phys. Rev. Lett. 2014;112:023202.
[90]   Coutinho, LH, Cavasso-Filho, RL, Rocha, TCR, Homem, MGP, Figueira, DSL, Fonseca, PT,
       Cruz, FC, de Brito, AN. Relativistic and interchannel coupling effects in photoionization and
       angular distributions by synchrotron spectroscopy of laser cooled atoms. Phys. Rev. Lett.
       2004;93:183001.

[91] Pengel D, Kerbstadt S, Johannmeyer D, Englert L, Bayer T, Wollenhaupt M. Phys. Rev. Lett. 2017;118:053003.

[92] Behr JA, Gwinner G. J. Phys. G: Nucl. Part Phys. 2009;36:033101.

[93] Smith Peter F. arXiv:1607.06876 [physics.ins-det].

Nicolas Sisourat and Alain Dubois
7 Semiclassical close-coupling approaches

1 Introduction

In ion-atom collisions, a very large number of inelastic channels (excitation, ionization, and electron capture or transfer, see equations (1.1a)–(1.1c) in Chapter 1) are generally open by energy conservation considerations, i. e.,

$$T_f = T_i + Q_{if} \tag{7.1}$$

where T_i (T_f) is the kinetic energy in the center-of-mass system for the initial (final) channel and Q_{if} is the internal energy change of the two partners during the collision. This is true as far as the translational energy for the incoming particles exceeds few electron-volts, corresponding to relative velocities larger than 0.01 a. u.[1] However the "probability" for a given process to occur depends spectacularly on the energy. Figure 7.1 shows the typical shapes[2] of the cross-sections as function of impact energy: they all present a bell-shape behavior, with capture dominating at low energies while excitation and ionization take over at large ones. In between, the cross-sections for the three types of processes have the largest values and are of comparable importance. In this range of energy, the so-called intermediate energy domain, the processes are strongly coupled and cannot be described independently of each other. It corresponds to values of the perturbation or Sommerfeld parameter defined in the introduction $\eta = \frac{Q}{v_p} \approx 1$. Note that in this domain where electron transfer is likely the magnitude of the collisional velocity is of same order than Bohr's velocity of the active electron in the initial state of the system (i. e., prior the scattering process).

This chapter is devoted to the theoretical description of ion-atom collisions in the intermediate energy domain where the electronic processes are coupled. We shall focus exclusively on collisions involving outer-valence electrons of neutral atoms and low charged ions. In this case, the intermediate energy range corresponds roughly to collision energies between 1 and 100 keV/u. In this energy range, the theoretical description of the scattering system should therefore be based on nonperturbative, also

---

[1] 0.01 a. u. $\approx 2.2 \times 10^4$ m s^{-1}. In the following, atomic units (a. u.) are used, unless otherwise stated.
[2] Depending on the systems, the cross-sections may present shoulders, several maxima, or oscillations.

Acknowledgement: This work has been financially supported by the ANR within the Investissements d'Avenir programme under reference ANR-11-IDEX-0004-02 (Labex Plas@par).

Nicolas Sisourat, Alain Dubois, Sorbonne Université, CNRS, Laboratoire de Chimie Physique-Matière et Rayonnement, F-75005 Paris, France, e-mails: nicolas.sisourat@upmc.fr, alain.dubois@sorbonne-universite.fr

https://doi.org/10.1515/9783110580297-007

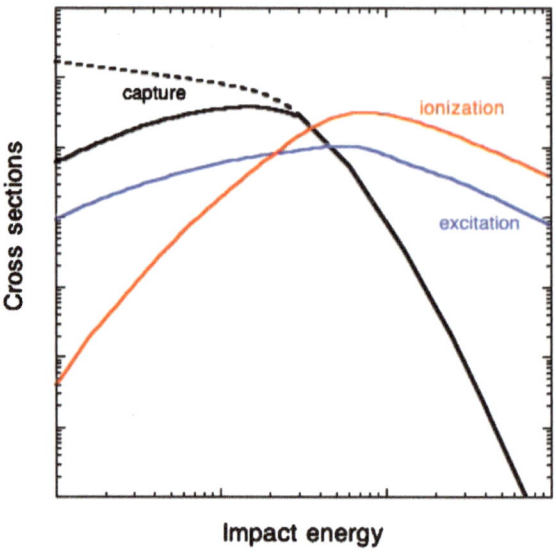

Figure 7.1: Cross-sections of electron capture, excitation, and ionization as function of impact energy. The black dashed line correspond to systems for which resonant capture is possible and dominates at low energies, e. g., $H^+ + H(1s) \rightarrow H(1s) + H^+$. *Note that the axes are not labeled to stay system independent.*

known as close-coupling, treatments, within the semiclassical approach: the internuclear target-projectile coordinate is then described classically, introducing the time variable (a clock; see, e. g., [1]) in the Hamiltonian operator of the scattering system. This is the starting point of the impact parameter approach where (i) the nuclei[3] relative motion is described by trajectories, generally safely reduced to constant velocity, straight-line ones, and (ii) a time-dependent quantum treatment is adopted for the electron dynamics.

In the following, we shall consider the study of the electronic dynamics in this context, where the time-dependent Schrödinger equation is solved using an asymptotic representation of the scattering system, i. e., using the electronic states of the isolated target and projectile.[4] We shall further restrict the focus of the chapter on the computations of total cross-sections. While differential cross-sections can be obtained

---

3 Here, we use the word "nuclei" for simplicity: in a frozen core electron approximation where only a given—small—number of valence electrons are considered active during the collision, the heavy particles are composed by the nuclei of Z protons and n_c frozen core electrons.

4 Quasi-molecular representations where the dynamics is expressed using the electronic states of the transient molecule formed by the two collision partners have also been developed and hold in the lower part of the intermediate energy range. This approach is very useful but it becomes conceptually and numerically less advantageous for higher energies. We shall not deal with the molecular representation in the following.

within the impact parameter method, the latter has been mostly employed to provide total cross-sections.

As pointed out in Chapter 1, atomic collisions are the optimal "laboratories" for the study of many-body quantum systems and of their dynamics. This fundamental point of view has been the main driving force which guides the efforts of thousands of researchers—including the authors of this book—since the beginning of the twentieth century. However, it is important to underline the relevance of the study of ion-atom collisions in the intermediate energy range. We therefore present below three important applications concerning other domains of physical and of societal importance that we are interested in:

1. Atomic and molecular collisions are highly important in astrophysics. For example, the solar wind is a hot plasma, composed of charged particles, electrons and ions (around 92 % of protons, 8 % of helium ions, and 0.1 % of heavier elements), that may collide with atoms and molecules (H, O, OH, CO, CH_4, NH_3, H_2O, ...) from solar-system bodies (planets and comets). X-ray emission induced by these collisions [2, 3, 4, 5, 6] can be detected and used to determine the structure and dynamics of the solar wind plasma, as well as of its environment (magnetic field). In particular, ions from the solar wind can capture electrons from planetary and cometary atoms and molecules into highly-excited electronic states which then deexcite radiatively emitting the X-ray photons [7]. This scenario, called the Solar Wind Charge-Exchange (SWCX) mechanism is the most probable process for the X-ray emission [8, 9].

2. In hadrontherapy, ion beams are used for tumor treatment. Ions used in such therapies have initially much more kinetic energies that what we are interested in this chapter. However, in the course of the beam through matter the ions lose gradually energy, mostly by ionizing the medium. It turns out that they deposit most of their energy, and thus create most of the radiation damage, in the high keV energy range which corresponds to the so-called Bragg peak [11]. Knowledge of the corresponding cross-sections would improve the modeling of ion-biological matter interactions, as stated in [10]: *the poor general understanding of the processes [...] like initial physical and chemical damage induction and their associated repair processes impedes the preferential mechanistic modeling of the biological response.*

3. Accurate cross-sections of the electronic processes occurring in the course of keV ion-atom collisions are required for the modeling and diagnostic of fusion plasmas. In particular, injection of hydrogen and helium beams into fusion reactors is a standard method to heat the plasma but also to probe it (e. g., temperature, composition, ...) via spectroscopic diagnostic tools: (i) spectroscopy of the neutral beams excited by the interactions with the plasma (so-called neutral beam spectroscopy [12, 13]), and (ii) spectroscopy of the plasma impurities after interacting with the neutral atoms of the beam (charge exchange spectroscopy [14]). Accurate cross-sections for the relevant electronic processes in ion-atom

collisions are therefore needed for modeling the beam penetration and the spectroscopic lines observed for the diagnosis of fusion plasmas; see, for example, [15].

The present chapter is composed of six sections to present step by step the theoretical description of electron dynamics in keV ion-atom collisions. In Section 2, we present the derivation of the impact parameter method, which leads to the eikonal equation, equivalent to a time-dependent Schrödinger equation (TDSE), for the electron(s) in the moving field of the nuclei. Contrary to the high energy regime which can be treated by time-dependent perturbative approaches, close-coupling techniques must be adopted to solve the TDSE in the intermediate energy domain. The nonperturbative approach based on finite Hilbert basis set representation is derived in a general way in Section 3. It is followed by a brief presentation of several close-coupling schemes available in the modern literature and which differ mainly by the choice of the basis sets employed (Section 4). We finally present in Section 5 some illustrative examples, focusing on convergence of the intensive computations required by such methods. As a conclusion of this chapter, we give in Section 6 some guidelines that theoretical investigations of keV ion-atom collisions might take in the future.

2 The impact parameter method

Considering an ion-atom collision composed of two heavy particles, target (T) and projectile (P) nuclei,[5] plus n_e electrons, the Hamiltonian of the system can be written in the laboratory frame as

$$\hat{H}^{\text{lab}} = \hat{T}^{\text{lab}} + \hat{V} \tag{7.2}$$

$$\hat{T}^{\text{lab}} = -\frac{1}{2M_T}\nabla^2_{R_T} - \frac{1}{2M_P}\nabla^2_{R_P} - \frac{1}{2}\sum_{i=1}^{n_e}\nabla^2_{R_i} \tag{7.3}$$

$$\hat{V} = \sum_{i=1}^{n_e} V_{Ti} + V_{Pi} + \sum_{i=1}^{n_e}\sum_{j>i}^{n_e} V_{ij} + V_{TP} \tag{7.4}$$

where the position vectors $\vec{R}_T$, $\vec{R}_P$, and $\vec{R}_i$ stand for the nuclei (of masses M_T and M_P) and electron (of mass $m_e = 1$ a. u.), respectively. The potential energy terms V_{TP}, V_{ij}, V_{Ti}, and V_{Pi} represent the Coulombic potentials between the nuclei, the electrons and

---

5 The distinction of the two heavy particles as target and projectile is done only to remind the experimental conditions but does not affect the generality of the following equations.

between the nuclei and the electrons, respectively.[6] Then an exact theoretical treatment requires to solve the full time-independent Schrödinger equation:

$$\hat{H}^{\text{lab}} \Psi_{\text{sys}} = E^{\text{lab}} \Psi_{\text{sys}} \tag{7.5}$$

where E^{lab} is the total energy of the system in the laboratory frame. Introducing the center of mass coordinate $\vec{R}_G$, the nuclear relative position vector $\vec{R} = \vec{R}_P - \vec{R}_T$ and $\vec{r}_i$ the position of the electrons with respect to an arbitrary point located on the internuclear vector, the Hamiltonian can be written as

$$\hat{H}^{\text{lab}} = -\frac{1}{2} \sum_{i=1}^{n_e} \nabla_{r_i}^2 - \frac{1}{2\mu_{TP}} \nabla_R^2 - \frac{1}{2M_{\text{tot}}} \nabla_{R_G}^2 + \hat{V} \tag{7.6}$$

with

$$\mu_{TP} = \frac{M_T M_P}{M_T + M_P} \quad \text{and} \quad M_{\text{tot}} = M_T + M_P \tag{7.7}$$

In equation (7.6), the infinite nuclear mass approximation (i. e., $M_P, M_T \gg m_e$) is used, with the neglect of the mass polarization terms, i. e., the crossed terms $\frac{1}{M_{\text{tot}}} \vec{\nabla}_{r_i} \cdot \vec{\nabla}_{r_j}$ [18].

The kinetic energy operator $\hat{T}^{\text{lab}}$ can be written as

$$\hat{T}^{\text{lab}} = \hat{T} - \frac{1}{2M_{\text{tot}}} \nabla_{R_G}^2 \tag{7.8}$$

where $\hat{T}$ corresponds to the internal kinetic energy. Since the potential energy terms depend only of the internal coordinates of the system, one can split the full Hamiltonian into two contributions

$$\hat{H}^{\text{lab}} = \hat{H} - \frac{1}{2M_{\text{tot}}} \nabla_{R_G}^2 \tag{7.9}$$

so that E^{lab} and Ψ_{sys} can be written as

$$E^{\text{lab}} = E + \frac{k_G^2}{2M_{\text{tot}}} \tag{7.10}$$

$$\Psi_{\text{sys}} = \Psi_{\text{int}} e^{i\vec{k}_G \cdot \vec{R}_G} \tag{7.11}$$

with

$$\hat{H} \Psi_{\text{int}} = E \Psi_{\text{int}}. \tag{7.12}$$

---

6 In frozen core approximations which have been extensively used, the terms V_{Ti}, V_{Pi} and V_{TP} are expressed as model potentials, describing the nucleus charge and the passive (core) electrons as a unique particle interacting with the active electrons of the system.

The internal Hamiltonian $\hat{H}$ reads

$$\hat{H} = \hat{H}_{el} - \frac{1}{2\mu_{TP}}\nabla_R^2 \tag{7.13}$$

where $\hat{H}_{el}$ is the electronic Hamiltonian

$$\hat{H}_{el} = -\frac{1}{2}\sum_{i=1}^{n_e}\nabla_{r_i}^2 + \hat{V}. \tag{7.14}$$

For ion-neutral collisional systems considered in the intermediate energy range, the dynamics of the projectile with respect to the target can be safely considered as weakly affected by the electronic dynamics. It is therefore reasonable to factorize out the main contribution of the nuclear dynamics from the total wave function Ψ_{int} as

$$\Psi_{int} = \Psi(\vec{R},\{\vec{r}\})\Xi(\vec{R}) \tag{7.15}$$

with the nuclear wave function $\Xi(\vec{R})$ written as a plane-wave [19]

$$\Xi(\vec{R}) = e^{i\vec{k}.\vec{R}} \tag{7.16}$$

where the wave vector $\vec{k}$ is defined by

$$|\vec{k}| = \sqrt{2\mu_{TP}E} \tag{7.17}$$

Note that the change of internal energy Q_{if} of the partners of the collision (due to electronic excitation, ionization, or transfer) is much smaller than the collision energy so that the modulus of wave vector is hardly affected in the final channels of the collision.

Using equations (7.15–7.17) and neglecting $\nabla_R^2\Psi(\vec{R},\{\vec{r}_i\})$ in front of the crossed term $\vec{\nabla}_R\Psi(\vec{R},\{\vec{r}_i\})$,[7] equation (7.12) reads

$$\left[\hat{H}_{el} - \frac{i}{\mu_{TP}}\vec{k}.\vec{\nabla}_R\right]\Psi = 0 \tag{7.18}$$

From there, one can introduce the semiclassical approximation in which it is assumed that the relative motion of the nuclei $\vec{R}$ is described by a classical trajectory $\vec{R}(t)$. Moreover, for the collision energies considered here, the projectile is scattered dominantly in the forward direction at very small angles. It is therefore reasonable to further assume that $\vec{R}(t)$ describes straight-line constant velocity trajectories[8]

$$\vec{R}(t) = \vec{b} + \vec{v}_p t \tag{7.19}$$

---

7 This approximation and the separation equation (7.15) are somehow equivalent to the assumptions used in the widely known Born–Oppenheimer approximation of molecular physics and theoretical chemistry [16, 17].

8 Note that only few studies have been performed using curved trajectories; see, e. g., [30, 31, 32].

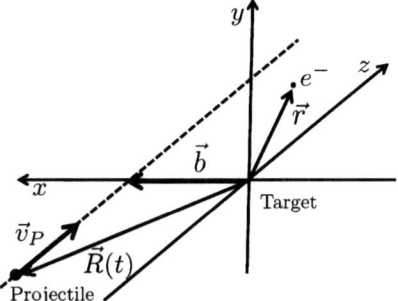

Figure 7.2: Collision geometry for a one-electron system. The impact parameter $\vec{b}$ and the projectile velocity $\vec{v}_P$ define the collision plane. The position of the electron with respect to the target center is denoted $\vec{r}$.

where $\vec{b}$ and $\vec{v}_P$ are the impact parameter and velocity, respectively; see Figure 7.2. The latter relates to $\vec{k}$ introduced in the plane-wave (equation (7.16)) describing quantally the nuclear motion

$$\vec{v}_P = \frac{1}{\mu_{TP}}\vec{k} \tag{7.20}$$

Using this relationship, equation (7.18) reads

$$i\hbar\frac{\partial\Psi}{\partial t} = H_{el}\Psi \tag{7.21}$$

which is known as the eikonal equation. The latter shows the same form than the time-dependent Schrödinger equation describing the electron dynamics in the moving field of the nuclei, in the straight-line constant velocity trajectory approximation. Note finally that this latter seems counter intuitive for the evaluation of differential cross-sections as a function of scattering angle, e. g., [27]. This is not the case and there exists an approximate but generally valid method to derive differential quantities from the results of semiclassical calculations, see [28, 29] and references therein. From a practical point of view, the straight-line constant velocity trajectory approximation implies only that the target-projectile scattering will not affect significantly the electronic dynamics obtained by the TDSE in the important region where the two collision partners are at the vicinity of each other. Note that though the approach proposed in this chapter is widely used, the modeling of the relative nuclear motion is not unique, the latter can be described classically by Coulombic trajectories and quantally by wave-packets, as it is illustrated in Chapter 4 of this book.

The TDSE equation (7.21) cannot be solved analytically even for the simplest collision systems (e. g., proton-hydrogen). A full numerical solution (spatial discretization) for systems having a unique active electron is now possible, e. g., [20, 21, 22, 23], given the tremendous progress in computational technologies. However, in the following we present more general approaches in which the TDSE is solved by expressing the electronic wave function onto finite Hilbert basis sets (state discretization). This type of approaches is sketched in the next two sections.

3 Close-coupling approach

In a close-coupling description, the electron dynamics is constrained to a configuration space defined by a finite set of basis functions $\chi_k(\{\vec{r}\})$.[9] The time-dependent electronic wave function is approximated by the following expansion:

$$\Psi(\{\vec{r}\}, t) = \sum_{k=1}^{N} a_k(t)\chi_k(\{\vec{r}\}) \tag{7.22}$$

If the set of basis functions $\chi_k(\{\vec{r}\})$ is complete, an exact solution is then obtained. For obvious numerical reasons, the expansion must be truncated and the configuration space is thus not complete. It is therefore important to choose wisely the basis functions used in the calculations. This leads to a number of different schemes. We present the most currently employed ones in Section 4.

Substituting the wave function ansatz (equation (7.22)) into equation (7.21) and projecting into each of the basis functions leads to a set of coupled first-order differential equations for the expansion coefficients. This set of equations can be written in a matrix form as

$$i\mathbf{S}\frac{d}{dt}\mathbf{a} = \mathbf{M}\mathbf{a} \tag{7.23}$$

where $\mathbf{a}$ is a vector containing the expansion coefficients and $\mathbf{S}$ and $\mathbf{M}$ are the time-dependent overlap and coupling matrices, respectively. The exact expression of these two matrices depends on the choice of the basis sets and cannot be written in a general form. Solving the TDSE with given initial conditions, i. e., given $\mathbf{a}(t \to -\infty)$, one can obtain the expansion coefficients after the scattering process, $\mathbf{a}(t \to +\infty)$, so that the wave function is known and can be analyzed in order to evaluate cross-sections. Note that various algorithms have been developed and implemented to solve numerically equation (7.23) but such discussion is outside the scope of this chapter.

We recall here that the electronic Hamiltonian depends on time due to the classical relative motion of the nuclei $\vec{R}(t)$ which depends explicitly on the impact parameter $\vec{b}$. For a given collision energy, equation (7.23) must therefore be solved for a set

---

9 It should be mentioned that is also possible, and sometimes numerically advantageous, to employ time-dependent basis functions [24, 25, 26]. However, in the following we restrict the presentation to the case where the time-dependence of the electronic wave function is solely included in the expansion coefficients.

of impact parameters that reproduces a given experiment. Assuming an uniform and cylindrical ion beam, the total cross-section for an inelastic process leading to a final state $\phi_f(\{\vec{r}\})$ is given by

$$\sigma_f = 2\pi \int_0^\infty b P_f(b)\, db \tag{7.24}$$

where the transition probability $P_f(b)$ reads

$$P_f(b) = \lim_{t \to \infty} |\langle \phi_f | \Psi(t) \rangle|^2. \tag{7.25}$$

In the latter equation, we have used the Dirac *bra-ket* notation for simplicity.

Within the impact parameter method, the above derivation is general. In the following, we review the different *implementations*[10] that have been proposed in the last few decades (see also [33]) and that are the most widely used nowadays.

4 Choice of the basis sets

Since the development of semiclassical close-coupling approaches, many schemes have been proposed. These methods can be divided in two major groups. In the first one, the functions describe the electronic states of the transient molecule formed during the collision. As mentioned in Section 1, we do not treat here these *molecular* approaches; a review of these methods can found in [34]. The second group corresponds to an asymptotic *atomic* picture: the basis functions are chosen to represent well the electronic states of the isolated target and projectile species. This class of methods is sometimes known as atomic-orbital close-coupling (AOCC).[11] The second group of methods can be further sorted into one-center and two-center approaches. In the former case, the electronic wave function is expanded into the electronic states of a single partner, generally the target species. This is particularly useful in the high collision energy range or for antiproton projectiles (see Chapter 3) where electron capture channels are weak or closed, respectively. In most recent research works, two-center approaches are generally adopted since they are better suited to describe electron capture: the electronic wave function is expanded into the electronic states of the target and the projectile. However, since the relative motion of the target and the projectile

---

10 We mean the choice of the basis set employed in each method and not the numerical techniques used to compute the necessary matrix elements and solving the set of coupled differential equations.
11 Note that this name may be misleading since these methods can in principle be used to simulate (i) ion-atom collisions involving several electrons where the use of orbitals is an approximation and (ii) ion-molecule collisions where the target states are then molecular states.

is treated classically, the electronic states of one partner moves in time with respect to the states of the other partner. One is forced to augment the atomic states by multiplicative phase factors, the Electron Translation Factors (ETFs), generally expressed as plane-waves, to take into account the relative translation between the target and projectile and ensuring the Galilean invariance of the results. A thorough discussion of ETFs can be found in [19]. We simply mention here that the ETFs make the computations of the overlap and coupling matrices (equation (7.23)) more involved, but do not lead to any changes in the derivation presented above.

An exhaustive list of the methods proposed in the past would be difficult to provide and in fact would not be very useful. Instead we focus on the methods that have stood the test of time. We therefore limit the discussion to the basis sets that have been employed in many studies and that are still currently in use. It should be also noted that most of these methods have been implemented for single-active electron systems and that their implementation differs mainly in the underlying basis sets used to construct the target and projectile electronic states.

4.1 Target and projectile exact eigenstates

For one-electron systems, the most straightforward implementation is to use the exact target and projectile eigenstates of negative energies, the latter being known analytically. This approach was used by several groups (see [33, 35, 36] and references therein). The exact eigenstates centred on the target and on the projectile allow a proper description of excitation and electron capture, respectively. However, only target and projectile bound states are included and such approach neglects completely the ionization process. As discussed above, in the intermediate energy range the three inelastic processes may be strongly coupled, and thus neglecting one of them may have a significant impact on the description of the other ones. It may lead to non-converged cross-sections, especially in the medium and high part of the intermediate energy range. A second drawback of this approach comes from the expression of the hydrogenic orbitals: they can be written as sums of Slater-type orbitals which have not a convenient form when multiplied by ETF in order to efficiently evaluate numerically the overlap and coupling matrix elements. Note that this approach can only be used for pure one-electron systems.

4.2 Target and projectile exact eigenstates with pseudostates

A clear improvement of the previous approach is the inclusion of so-called pseudostates. The latter improve the convergence of the calculations with respect to the basis of exact eigenstates in two ways. First, they can approximately describe excitation and electron capture into Rydberg states which cannot be included explicitly in the set of exact low-lying eigenstates. Second, the pseudostates of positive energies can provide a discretized electron continuum, and thus be used to describe ionization in an approximate way. Several implementations of this general idea have been proposed and differ by the type of pseudostates employed. Coulomb Sturmian functions which are scaled hydrogen-like wave functions satisfying different orthonormalization conditions were proposed in [37] and extended in the last three decades by Winter [38, 39]. Orthogonal Laguerre basis are used in the convergent close coupling method developed by a group in Australia (see [40] for a recent review). States obtained by diagonalization of the target and projectile Hamiltonian matrices in Slater-type orbital basis sets (with hydrogen-like exponents, i. e., nuclear charge over principal quantum number, Z/n) have been used in [41, 42, 43, 44]. A particularly appealing way to construct pseudostates that are, in principle, optimal for the description of the electronic dynamics is the so-called basis generator method (BGM) [45]. In the latter, the pseudostates are constructed by repeated applications of the projectile interaction operator, thus minimizing the configuration space that cannot be covered by the truncated wave function expansion.

4.3 Target and projectile pseudostates

The former approaches include directly the exact eigenstates or can obtain exactly these latters through a proper choice of the basis set parameters (exponents, order of the polynomials). One can also employ only pseudostates in the wave function expansion.[12] Obviously, some of these pseudostates should provide an excellent description of the target and projectile bound states in order to include properly initial, excitation, and electron capture channels in the calculations. A general approach in this case is to diagonalize the electronic Hamiltonian matrix of the isolated collision partners in a chosen basis set.

Gaussian-type orbitals (GTOs) are now widely used in this context, as they are in quantum chemistry. They allow indeed efficient computation of the overlap and coupling matrix elements, especially when ETFs are included, and provide a fairly good description of the exact bound states.[13] Furthermore, due to the addition theorem, the use of GTOs allows to extend fairly easily the computations of the overlap and coupling matrix elements to multielectron systems and ion-molecule collisions [48]. GTOs have been employed in, for example, [49, 50, 51]. We further discuss their use in the next section.

---

12 This scheme is unavoidable when considering multiactive-electron atoms but also for quasi-one electron systems, e. g., atomic systems such alkaline atoms for which the binding of the valence electron may be described by model potentials.
13 Note that however GTOs cannot reproduce the cusp at the nucleus of the true eigenstates (see [46] for a detailed discussion.)

4.4 One-center expansions

While two-center expansions are mostly used nowadays in order to describe appropriately electron capture, we mention here that there are recent works where one-center expansions have been employed. The latter are usually easier to implement since the inclusion of ETFs is not required and the computations of the overlap and coupling matrix elements can be done efficiently. It should also be noted that despite the one-center character of the expansion the electron capture can be approximately treated through the use of pseudocontinuum states. One-center expansions have been recently used in, e. g., [52, 53, 54].

5 An illustrative example

In this section, we present some of our recent results in order to show the different challenges encountered in the use of semiclassical close-coupling approaches. The collisional system Li^{3+}-H is used for illustration. It is a true three-body system (i. e., one-electron system) which allows in many cases an almost numerically exact solution. Furthermore, it is of practical importance in fusion energy research, through the modeling and diagnosis of plasma [12, 13, 14].

In our approach, the target and projectile states (see equation (7.22)) are described as linear combinations of N_g Gaussian-type orbitals

$$\mathcal{G}_i(\vec{r}) = \mathcal{N}_i x^{u_i} y^{v_i} z^{w_i} e^{-\alpha_i r^2} \tag{7.26}$$

where $\mathcal{N}_i$ is a normalization factor and u_i, v_i, and w_i are integers which allows the description of the spherical harmonics in Cartesian coordinates. The choice of the $\mathcal{G}_i$ functions to include into the basis set is crucial since it determines (i) the Hilbert space spanned by the expansion and (ii) how accurate the target and projectile states are described. First, it is common practice to chose all possible combinations of u_i, v_i, and w_i up to a given $\ell = u_i + v_i + w_i$, allowing the building-up of all spherical harmonics $Y_{\ell,m}$. Second, the choice of the GTO exponents α_i should be performed to have a correct description of the states important in the scattering process. In our recent works, we have used even-tempered basis sets which ensures an even coverage of the Hilbert space [47]. This is indeed particularly important to describe equally excitation, ionization as well as electron capture. The exponents then follow a geometric sequence according to

$$\alpha_i = \alpha/\beta^{i-1} \tag{7.27}$$

with $\beta > 1$ and $i = 1, \ldots, N_g$. Using this geometric sequence, only two parameters need to be chosen to best describe the electronic states of the isolated partners: for given values of (α, β), the increase of the number N_g of GTO allows to include more diffuse orbitals, and, consequently, higher energy states. Alternatively, for given values of α and of the smallest exponent α_i, the increase of N_g (i. e., decrease of β) improves the quality of the states, keeping approximatively the size of the Hilbert space spanned by the basis set. Thus these schemes provide systematic ways to investigate the convergence of the cross-sections with respect to the basis sets.

Given a GTO basis set, the target and the projectile states are obtained by diagonalizing the corresponding atomic Hamiltonian matrix. As explained above, these states are sometimes called pseudostates since they are not, strictly speaking, exact eigenstates of the atomic species considered. Indeed from this diagonalization procedure, one can obtain eigenvalues which are very close to the energies of the *exact* states (the lowest-lying ones). It also provides some pseudostates with negative energies which does not correspond to real states as well as some with positive energies, used in the calculations to take into account ionization. We mention here that these states are also sometimes named pseudostates in the literature in order to differentiate them from states corresponding to negative energies (i. e., bound states). To avoid any confusion, we will call any state obtained through diagonalization pseudostates and among these pseudostates, those corresponding to positive energies will be named pseudocontinuum states.

The first step to perform in an actual calculation is therefore the diagonalization of the electronic Hamiltonian matrix of the target on one hand and of the projectile on the other hand. We show in Table 7.1 the energies of the first bound pseudostates obtained by diagonalizing the electronic Hamiltonian matrix of the hydrogen atom using 5 even-tempered basis sets of increasing quality (from B1 to B5). The exponents are reported in the Appendix (Section 7).

Table 7.1: Energy (in atomic unit) of the hydrogen pseudostates obtained with the 5 different basis sets considered in this example. The last column provides the exact energies.

Orbitals	B1	B2	B3	B4	B5	exact
1s	−0.489569	−0.499032	−0.499932	−0.499982	−0.499984	−0.500000
2s	−0.119491	−0.124986	−0.124988	−0.124997	−0.124998	−0.125000
2p	−0.124985	−0.124523	−0.124977	−0.124989	−0.124990	−0.125000
3s	−0.051387	−0.054698	−0.055489	−0.055544	−0.055539	−0.055556
3p	−0.054668	−0.054818	−0.054959	−0.055096	−0.055198	−0.055556
3d	−0.053939	−0.055406	−0.055546	−0.055554	−0.055553	−0.055556

Table 7.1 shows that the description of the pseudostates energies is systematically improved from B1 to B5, from about 2 % differences with respect to the exact values to $< 10^{-3}$ % when increasing by a factor about two the number of GTOs for each angular momentum. Note that the Li^{2+} states are obtained with similar precision when ex-

pressed on equivalent GTO basis sets, with scaled exponents compared to the hydrogen ones; see Section 7.

In the following, we report the total cross-section for electron capture, excitation, and ionization computed with these five basis sets. Note that all target and projectile pseudostates (with negative and positive energies) produced by diagonalization step are included in the electronic wave function (equation (7.22)): on each center, from 47 states for B1 to 83 for B5. We discuss the effect of neglecting some of these pseudostates later in this chapter.

The cross-sections for electron capture, excitation, and ionization in Li^{3+}-H collisions are shown in Figures 7.3, 7.4, and 7.5, respectively.[14] We first discuss the results for the electron capture processes (Figure 7.3). The cross-sections exhibit a maximum at a collision energy of about 20 keV/u, which corresponds to impact velocity $v_p \approx 0.9$ a. u. close to the *Bohr velocity* of the electron in the first orbit. The comparison between the results obtained with the different basis sets shows that the B1 basis set is insufficient to properly described the electron capture: the cross-sections are too large around 20 keV/u and sligthly too small at larger collision energies. By improving the quality of the basis sets, one sees that the cross-sections do not change significantly, demonstrating that the electron capture cross-sections obtained with the B4 and B5 basis sets are converged (for all practical purposes) with respect to the basis set.

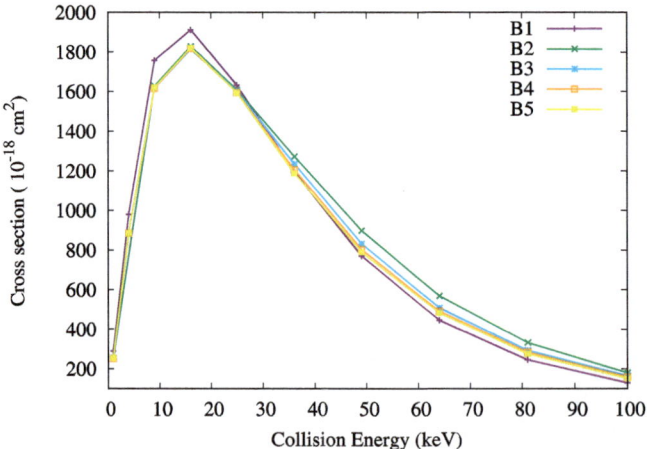

Figure 7.3: Total cross-sections for electron capture in Li^{3+}-H computed with 5 different basis sets (see text).

The choice of the basis set is even more important for the excitation and ionization processes in this collision system, as shown in Figures 7.4 and 7.5. The excitation cross-

---

14 Note that unlike the usual use of log-log scale, our cross-sections are displayed on linear scales.

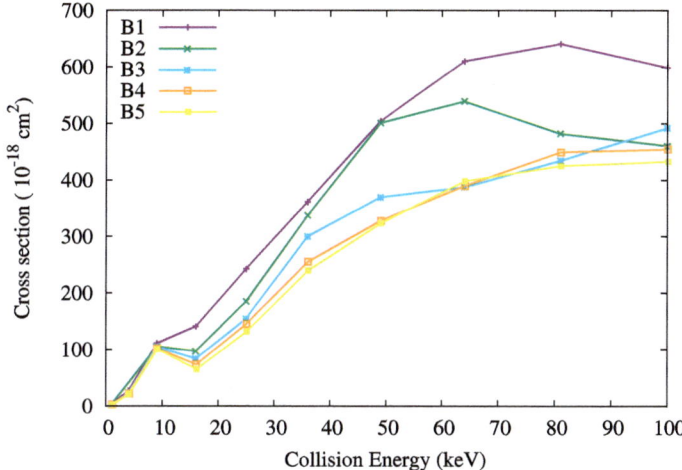

Figure 7.4: Total cross-sections for excitation in Li^{3+}-H computed with 5 different basis sets (see text).

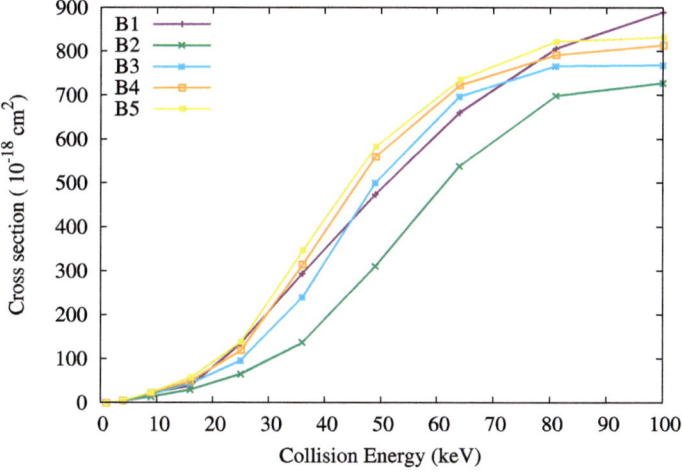

Figure 7.5: Total cross-sections for ionization in Li^{3+}-H computed with 5 different basis sets (see text).

sections increase from 1 to 100 keV/u. The difference between the cross-sections obtained with B1 and B5 basis sets is up to 50 % and that between B3 and B5 basis sets are up to about 20 %, showing a slow convergence. The differences between the results of the two largest basis sets are however fairly small.

Similar conclusions can be drawn for the ionization cross-sections; see Figure 7.5. The latter also illustrate the *nonvariational* character of the close-coupling approaches, which is an important aspect of these methods: indeed, by improving the basis sets the ionization cross-sections do not converge monotonously in the complete energy range. For example, at 100 keV/u collision energy the cross-section obtained with the B1 basis set is higher than the converged one while that obtained with the B2 and B3 basis sets are lower. However, one can see a slow but systematic convergence when going from B2 to B5. In other words, calculations with a single basis set do not provide an upper or lower limit of the cross-sections and several basis sets should be employed to check the convergence of the cross-sections.

The previous discussion highlighted the importance of the GTO basis sets employed. The same issue applies to any methods that do not use the exact electronic states of the target and the projectile. For any systems with more than one-electron, this is always the case. A more common issue with all close-coupling methods (even those using exact electronic states) is the choice of the states (or pseudostates) to include in the electronic wave function expansion. Obviously, one has to truncate the sum in equation (7.22). It is therefore necessary to check that the missing part of the electronic configuration space does not play an important role in the description of the electron dynamics. This convergence check is generally done by comparing the cross-sections of interest computed with an ever increasing, when possible, number of states (or pseudostates) in the wave function expansion. To illustrate this point, we focus on the cross-sections for excitation into $H(n = 3)$. In the following, we report calculations where we employed a single GTO basis set (the converged B5 basis set) but we changed the number of states included in the close-coupling expansion. The truncations were chosen to approximately include (or neglect) some physical channels (e. g., electron capture or ionization).

The cross-sections computed with basis B5 but including in the wave function expansion either only bound states on both centers (i. e., neglecting ionization) or only states centered on the target (i. e., neglecting capture) are compared with the full two-center expansion in Figure 7.6. The cross-sections obtained with the wave function containing only bound states are close to that computed with the complete expansion at energies below 20 keV/u since ionization is not likely in this energy range. Above 20 keV/u, the latter becomes important and neglecting it leads to an overestimation of the excitation cross-sections. Conversely, the excitation cross-sections obtained with the one-center (target) expansion are very close to the converged ones at high collision energy where electron capture cross-sections are small and are enormously overestimated at the lower collision energy in which electron capture processes are substantial. It should be noted that the one-center expansion still includes approximately electron capture through the pseudo continuum states of the target. However, it is practically difficult (if not impossible) to obtain and include sufficient pseudocontinuum states in this type of expansion to reach convergence. This example illustrates clearly the importance of including both ionization and capture processes in the calculations, even when one is interested in the excitation processes.

We have shown an example of the convergence of close-coupling calculations, for the system and the energy domain under consideration. Comparing the results from

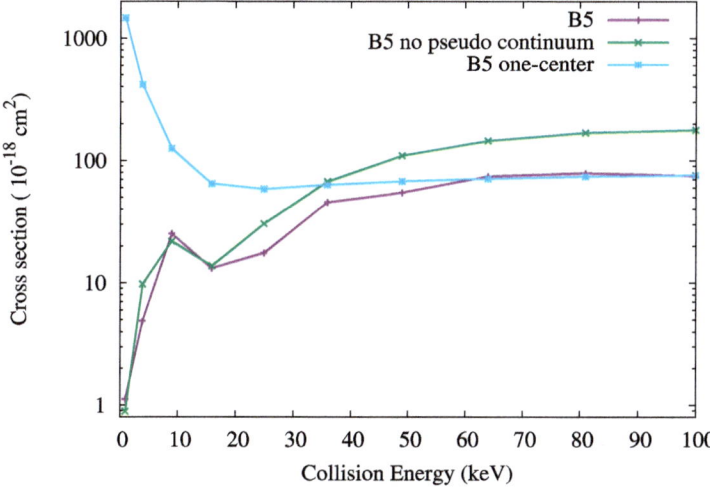

Figure 7.6: Cross-sections for excitation into $H(n = 3)$ in Li^{3+}-H computed with 3 different basis sets (see text).

the two largest basis, it can be shown that the convergence was in general less than few % and, at worse, 10 % for ionization and for some energies. The purpose of this chapter was only to illustrate this important challenge of close-coupling methods and not to perform the ever-best calculations. Then a reasonable question could be: can we reach a ‰ convergence by increasing the size of the basis set? The answer should unfortunately be hand-waving since, in absence of clear and general mathematical criteria, one cannot state it for sure. A priori one should improve the convergence when doing so. However, when using two-center pseudostates expansions, one can reach overcompleteness of the basis set and then linear dependencies can gradually show up, corrupting the quality of the results, at least for some impact energies. This is especially true when using large numbers of pseudostates of positive energy on both centers: the diagonalization of the target Hamiltonian matrix can accidentally produce a pseudostate with energy corresponding to electron velocity close to the impact velocity. This state may behave as a projectile state already included in the basis set. Then two elements of the basis set model the same real state and introduce therefore linear dependencies and numerical divergencies, e. g., when inverting the overlap matrix required to solve the set of coupled differential equations; cf. equation (7.23). This kind of instability was thoroughly discussed in the context of spurious oscillations observed for $H^+ - H$ collisions in cross-sections as function of impact energy; see [56, 57]. One-center basis sets do not have this drawback and, as mentioned before, require shorter computation times than two-center expansions. However their convergence is very slow, as far as capture processes are important.

Note that the calculations discussed above are partly related to one of our investigations. Here, the data are only reported for illustrative and pedagogical purposes.

A thorough discussion of the calculations and a comparison with other works are given in [51].

6 Conclusions and future directions

In this chapter, we have presented the basic equations of the impact parameter method and the subsequent semiclassical nonperturbative (i. e., close-coupling) approaches. Formula to compute the total cross-sections for excitation, ionization, and electron capture are given. We then discussed the different implementations that are currently employed by researchers, sometimes after decades of development. A few examples illustrating the different steps of a semiclassical close-coupling approach and a few important keypoints of such calculations were finally reported. The purpose of this chapter was not to be exhaustive and further details can be found in books and reviews, as B. H. Bransden and M. R. C. McDowell, 1992 [28] and W. Fritsch and C. D. Lin, 1991 [33].

Obviously, the features we chose to discuss reflect our own research and we left out some other important aspects. For example, we just briefly mentioned semiclassical non-perturbative *molecular* approaches, which are in fact widely used in our field. We did not discuss how to compute differential cross-sections within the impact parameter method. A proper presentation of such computations can be found, e. g., in [19, 55].

The main purpose of this book is to present all aspects of ion-atom collisions. An obvious extension is that of ion-molecule collisions. While the basic equations presented above are in principle applicable to ion-molecule collisions, the implementation can be quite different.[15] We mention here a few references (see also references therein) for a reader interested in these collisions [58, 59]. Talking about implementation, we did not discuss any numerical algorithms in this chapter. Matrix diagonalizers and differential equations solvers can be found in any numerical libraries.

To conclude, we think it is fair to state that the basic physics taking place in keV ion-atom collisions are now well understood. Of course, there are still (and probably will always be) some particular examples for which we do not understand the peculiar behavior of a given cross-section but we believe that this lack of understanding does not come from a missing piece of physics. One-electron systems have been thoroughly investigated and converged simulations have been achieved for most of them. This is not true for N active electrons systems ($N \geq 2$), where there is room for improving

---

15 In the intermediate energy region, the molecular nuclei can be safely considered as fixed in space as far as the scattering is concerned. When neglecting the vibration and rotation degrees of freedom, the only main difference between ion-atom and ion-molecule collisions stems from the multicenter features of the latter; see, e. g., [60, 61].

the currently available calculations. Such direction of research is vital to be pursued though one cannot expect that it provides important breakthroughs in the few-body problem.

It does not mean that atomic and molecular physicists are done with keV ion-atom collisions. It is our opinion that this field is just getting started. As mentioned in the Introduction of this chapter, inelastic processes in keV ion-atom collisions play a central role in many applications going from ion-based therapies to fusion plasma and astrophysics. Accurate cross-sections are crucial input data of more macroscopic models. However, to date there is no complete close-coupling calculations for most of the systems of interest. Let us take the example of a collision between a proton and a water molecule, which is the foremost important collisional system in the context of hadrontherapies. Several models have been proposed which either neglect all electrons but one, or the multicenter character of the water target, or both of them. While these models are helpful, they should be compared to more accurate calculations, when detailed experimental investigations are rare. We believe this is the path that should be taken in the future.

The comparison of the current theoretical state-of-the-art in keV ion-atom and ion-molecule collisions with standard quantum chemistry is striking. Quantum chemists can nowadays compute properties of hundreds of electronic states for polyatomic molecules having tens of electrons. There is no reason such level of development cannot be achieved in our field. Many efficient algorithms and tricks can directly be *borrowed* from quantum chemistry. Of course, this will require a substantial effort in implementing these techniques in codes. We hope that this chapter will be helpful to the young researchers ready for this challenge.

7 Appendix

The target and projectile states used in the Li^{3+}-H calculations were computed by diagonalization of the Hamiltonian matrix in a given Gaussian-type orbitals (GTO) basis set. We used Cartesian GTO defined as

$$\mathcal{G}_i(\vec{r}) = \mathcal{N}_i x^{u_i} y^{v_i} z^{w_i} e^{-\alpha_i r^2} \tag{7.28}$$

where $\mathcal{N}_i$ is a normalization factor and u_i, v_i, and w_i are integers. For a given value of angular momentum ℓ, we use all combinations of u_i, v_i, and w_i such $u_i + v_i + w_i = \ell$. The GTO exponents (α_i) follow a geometric sequence according to

$$\alpha_i = \alpha/\beta^{i-1} \tag{7.29}$$

with $\beta > 1$ and $i = 1, 2, \ldots, N_g$. The parameters and number (N_g) of GTOs used in the calculations for the hydrogen states are listed in Table 7.2. For the lithium states, the

Table 7.2: Parameters of the Gaussian-type orbitals (GTOs) used for the calculations reported in section 5: α and β define the exponents and N_g is the number of GTOs.

ℓ	B1			B2			B3			B4			B5		
	α	β	N_g	α	β	N_g	α	β	N_g	α	β	N_g	α	β	N_g
0	100.0	2.51	6	100.0	2.15	7	100.0	1.93	8	100.0	1.78	9	100.0	1.67	10
1	1.0	2.17	7	1.0	1.93	8	1.0	1.78	9	1.0	1.67	10	1.0	1.59	11
2	0.65	4.64	4	0.65	3.17	5	0.65	2.51	6	0.65	2.15	7	0.65	1.93	8

GTOs exponents were obtained according to the scaling law [62] $\alpha^Z = \alpha^H * Z^2$ where α^H are the exponents for the hydrogen atomic states and Z is the nuclear charge of the projectile.

Bibliography

[1]   Briggs JS. Phys. Rev. A. 2015;91:052119.
[2]   Bhardwaj A et al. Planetary and Space Science. 2007;55:1135.
[3]   Lisse CM et al. Science. 1996;274:205.
[4]   Dennerl K, Englhauser J, Truemper J. Science. 1997;277:1625.
[5]   Dennerl K. Astron. Astrophys. 2002;394:1119.
[6]   Dennerl K, Burwitz V, Englhauser J, Lisse C, Wolk S. Astron. Astrophys. 2002;386:319.
[7]   Cravens TE. Geophys. Res. Lett. 1997;24:105.
[8]   Cravens TE. Science. 2002;296:1042.
[9]   Gunell H, Holmstroem M, Kallio E, Janhunen P, Dennerl K. Adv. Space Res. 2005;36:2057.
[10]   Schardt D, Elsässer T, Schulz-Ertner D. Rev. Mod. Phys. 2010;82:383.
[11]   Belkić D, editor. Theory of Heavy Ion Collision Physics in Hadron Therapy. Adv. Quantum Chem. 65:1; 2013.
[12]   von Hellermann MG et al. Phys. Scr. 2005;120:19.
[13]   Delabie E et al. Plasma Phys. Control. Fusion. 2010;52:125008.
[14]   Andersonet H et al. Plasma Phys. Control. Fusion. 2010;42:781.
[15]   http://open.adas.ac.uk/.
[16]   Born M, Oppenheimer R. Ann. Phys. 1927;84:457.
[17]   Born M, Huang K. Dynamical Theory of Crystal Lattices. New York: Oxford University Press; 1954.
[18]   Brandsden BH, Joachain CJ. Physics of atoms and molecules. Prentice Hall; 2003.
[19]   McCarroll R. Charge exchange and ionization in ion-atom collision. Atomic and molecular collision theory NATO ASI Series B, New York: Plenum Publishing Corporation; 1982.
[20]   Bottcher C. Phys. Rev. Lett. 1982;48:85.
[21]   Riley ME, Ritchie AB. J. Phys. B: At. Mol. Opt. Phys. 1999;32:5279.
[22]   Pindzola MS, Schultz DR. Phys. Rev. A. 2008;77:014701.
[23]   Sælen L, Birkeland T, Sisourat N, Dubois A, Hansen JP. Phys. Rev. A. 2010;81:022718.
[24]   Beck MH, Jäckle A, Worth GA, Meyer H-D. Phys. Rep. 2000;324:1.
[25]   Meyer H-D, Worth GA, Gatti F, editors. Multidimensional Quantum Dynamics: MCTDH Theory and Applications. Weinheim: Wiley-VCH; 2009.
[26]   Kroneisen OJ, Lüdde HJ, Kirchner T, Dreizler RM. J. Phys. A: Math. Gen. 1999;32:2141.

[27] Dowek D et al. Phys. Rev. A. 1996;54:970.
[28] Bransden BH, McDowell MRC, Charge Exchange and the Theory of Ion-Atom Collisions. Oxford University Press; 1992.
[29] Dubois A, Nielsen SE, Hansen JP. J. Phys. B: At. Mol. Opt. Phys. 1993;26:705.
[30] Dickinson AS, Hardie DJW. J. Phys. B: At. Mol. Opt. Phys. 1979;12:4147.
[31] Fritsch W. J. Phys. B: At. Mol. Opt. Phys. 1982;15:L389.
[32] Forster C, Shingal R, Flower DR, Bransden BH, Dickinson AS. J. Phys. B: At. Mol. Opt. Phys. 1988;21:3941.
[33] Fritsch W, Lin CD. Phys. Rep. 1991;202:1.
[34] Errea LF et al. J. Phys. B: At. Mol. Opt. Phys. 1994;27:3603.
[35] Igenbergs K et al. J. Phys. B: At. Mol. Opt. Phys. 2009;42:235206.
[36] Wang J, Hansen JP, Dubois A. J. Phys. B: At. Mol. Opt. Phys. 2000;33:241.
[37] Shakeshaft R. Phys. Rev. A. 1978;18:1930.
[38] Winter TG. Phys. Rev. A. 1982;25:697.
[39] Winter TG. Phys. Rev. A. 2013;87:032704.
[40] Bray I et al. J. Phys. B: At. Mol. Opt. Phys. 2017;50:202001.
[41] Liu L, Li XY, Wang JG, Janev RK. Phys. Plasmas. 2014;21:062513.
[42] Igenbergs K et al. J. Phys. B: At. Mol. Opt. Phys. 2012;45:065203.
[43] Yan LL et al. Eur. Phys. J. D. 2015;69:26.
[44] Lu J, Hansen JP, Nielsen SE, Dubois A. J. Phys. B: At. Mol. Opt. Phys. 1998;31:3665.
[45] Zapukhlyak et al. J. Phys. B: At. Mol. Opt. Phys. 2005;38:2353.
[46] Klopper W, Kutzelnigg W. J. Mol. Struct. (THEOCHEM). 1986;135:339.
[47] Cherkes I, Klaiman S, Moiseyev N. Int. J. of Quant. Chem. 2009;109:2996.
[48] Sisourat N, Pilskog I, Dubois A. Phys. Rev. A. 2011;84:052722.
[49] Toshima N. Phys. Rev. A. 1999;59:1981.
[50] Labaigt G, Jorge A, Illescas C, Beroff K, Dubois A, Pons B, Chabot M. J. Phys. B: At. Mol. Opt. Phys. 2015;48:075201.
[51] Ibaaz A, Esteban Hernandez R, Dubois A, , Sisourat N. J. Phys. B: At. Mol. Opt. Phys. 2016;49:085202.
[52] Ford AL, Reading JF, Hall KA. J. Phys. B: At. Mol. Opt. Phys. 1993;26:4537.
[53] Martin F. J. Phys. B: At. Mol. Opt. Phys. 1999;32:501.
[54] Errea LF. Phys. Rev. A. 2006;74:012722.
[55] Briggs J, Macek JH. Adv. At. Mol. Opt. Phys. 1990;28:1.
[56] Kuang J, Lin CD. J. Phys. B: At. Mol. Opt. Phys. 1996;29:1207.
[57] Toshima N. J. Phys. B: At. Mol. Opt. Phys. 1997;30:L131.
[58] Lüdde HJ, Spranger T, Horbatsch M, Kirchner T. Phys. Rev. A. 2009;80:060702.
[59] Sisourat N, Pilskog I, Dubois A. Phys. Rev. A. 2011;84:052722.
[60] Errea LF, Gorfinkiel JD, Macías A, Méndez L, Riera A. J. Phys. B: At. Mol. Opt. Phys. 1997;30:3855.
[61] Caillat J, Dubois A, Sundvor I, Hansen JP. Phys. Rev. A. 2004;70:032715.
[62] Pye CC, Mercer CJ. J. Chem. Educ. 2012;89:1405.

Reinhold Schuch

8 Interference effects in electron capture collisions

1 Introduction

When ions interact with matter, electron capture can occur, i. e., the transfer of electrons from the atoms, molecules, or ions of matter to the impinging ions. The roots of the field date back to the middle of the nineteenth century with the study of phenomena in gas discharges. Early in the twentieth century, after the discovery of radioactivity, exploration of the penetration of charged particles in matter and their accompanying ionization phenomena became an active research area. N. Bohr showed a lifelong interest in the subject, often using it as an important test of treatments of atomic dynamics. Bohr's first paper [1] on the penetration of charged particles through matter was as early as in 1913 when he treated the slowing down of charged particles in matter on the basis of Rutherford's atomic model. And his last paper [2] was written together with Jens Lindhard, on electron capture and loss, published in 1954. The understanding of these atomic collision processes are closely connected to the development of quantum mechanics. Here, first attempts date back to 1928 [3] and 1930 [4]. And still today, there are theoretical attempts to reach a more complete description of the quantum dynamics of electron capture. With such a long history of developments, one may ask if surprises and developments are there still to be expected. The articles in this book and also the following review will provide ample evidence for many remarkable developments and the remaining challenges in this field.

Electron capture is an essential process in various scenarios of interacting matter, in astrophysics, in laboratories with ion beams and in fusion devices. It is very much occurring in astrophysical environments, where ions are abundant, such as stars, supernovae, solar plasma, planetary atmospheres, etc. In the laboratory, electron capture determines the charge of the ion and influences energy loss that is important for transport of ions in matter, thus for material probing and modifications, or lifetime of stored or trapped ions.

New frontiers originate often from new experimental opportunities with new instruments having higher resolution and sensitivities, an increase of the dynamic range to higher and lower energies, an increase of ion intensities, and consequently luminosity with new target technologies that allow to measure smaller and smaller

Note: Dedicated to Horst Schmidt-Böcking 80th birthday.
Acknowledgement: Valuable comments by Professors H. Schmidt and M. Schulz are acknowledged.

Reinhold Schuch, Physics Department, Stockholm University, AlbaNova, 10691 Stockholm, Sweden, e-mail: schuch@fysik.su.se

https://doi.org/10.1515/9783110580297-008

cross-sections. Fast and radiation resistant particle detectors with position resolution make angular resolved coincidence measurements of, e. g., radiation or electrons and/or recoil-ions from ion-atom collisions possible. A new experimental approach was also generated by the development of recoil-ion-momentum spectroscopy. For investigations of electron capture reactions, the invention of the reaction microscope [5], in which the momentum vectors of all outgoing particles (ions and electrons) are measured with high resolution and high efficiencies, represents a major break-through. For pure single electron capture processes, the corresponding experimental task was to measure the three momentum components for the recoiling target ions with high precision and high resolution in coincidence with the charge-changed ions. The COLTRIMS technique (cold target recoil-ion-momentum spectroscopy), a forerunner to reaction microscopy, is ideally suited for this task [6].

To start with, we consider only single-electron capture processes occurring in collisions of projectile ions X^{q+} with charge q with target atoms or molecules A: $X^{q+} + A \rightarrow X^{(q-1)+} + A^{+}$.

For describing the mechanisms of electron capture, it is useful to distinguish several velocity regimes of the ion impacting on matter. These come about through the different mechanisms governed by kinematics, momenta, and energetics of the electron transfer process. Those determine also validity ranges of approximations in the theoretical descriptions and the contributions from competing processes. A natural parameter of separating different ion velocity (v_i) regimes is the Bohr velocity (v_B) of the electron to be captured. At low v_i compared to v_B the momentum transfer and also the energy release ΔE in capture is negligible compared to the bound states momenta and energies. There a resonant transfer of the electron between the bound states dominates. The energetics of capture is illustrated in the schematic energy diagram of Figure 8.1. The kinetic energy of the ion causes a small energy shift T between the levels. But still electron capture is a complicated process for several reasons, such as the Coulomb interaction between two ions in the exit channel and its absence in a collision of an ion with an atom in the entrance channel, and more that are discussed throughout this article.

This illustration of the energy balances in Figure 8.1 gives therefore a quite incomplete picture of capture as it, e. g., does not visualize the momentum balances and other properties of the many body system. Capture works in quite different ways at low velocity of the ion as compared to when the ion is much faster (velocity v_i), then the typical Bohr velocity (v_B) of the electron to be captured. There are several reviews that elucidate the theoretical aspects in detail [7, 8, 9].

At high v_i compared to v_B ($v_i \gg v_B$), the energy release ΔE in capture can be facilitated by the emission of a photon. The photon emission is an effective way to get rid of the large kinetic energy and the cross-section of this radiative electron capture (REC) can become dominant. This is the analogous mechanism to the pick-up of a free electron (radiative recombination), to be distinguished from the radiative cap-

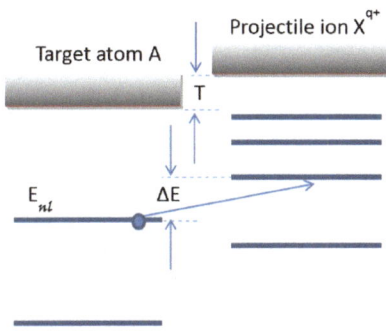

Figure 8.1: Schematic view of the energy balance for electron capture in a collision. The active electron (mass m) in target atom A in the state of principal and orbital quantum numbers n, l with the binding energy $E_A(n, l)$ is captured by the ion X^{q+} (velocity v) to the state $n_1 l_1$ and binding energy $E(n_1 l_1)$. The active electron gets a kinetic energy $T = m/2v^2$ moving with the resulting ion after the reaction. So it is $\Delta E = E_A(n, l) - E(n_1 l_1) + T$ the energy defect of the reaction (neglecting a small recoil energy of A^{1+}) that is taken from the kinetic energy of the ion.

ture of an electron bound in an atom. Another aspect of this process is considered below.

In this review, we concentrate on aspects of non-radiative electron capture where classical and quantum effects play striking roles, showing visible consequences in the differential cross-sections. For the message we like to transmit here, we stay mostly in a somewhat simplified picture of a single active electron. The article is organized as follows: Chapter 2 reveals classical and quantum behaviors in electron capture from atoms at high projectile velocity. In Chapter 3, we treat interference effects in capture from simple molecules at high velocity, and in Chapter 4, we show the appearance of interference effects in state selective capture in slow atomic collisions.

2 Classical and quantum behavior in electron capture from atoms at high projectile velocity

Electron capture at high velocities, i. e., high compared to v_B, is competing with the much more likely case of ionization, and its probability is relatively small. Descriptions of capture reactions is a rather tedious task, accompanied by a strong coupling of rearrangement channels with ionization. The electron in 1s of hydrogen has an average speed of $\alpha c = v_0$ (this is the velocity unit $v_0 = 1$ in atomic units a. u., where c is the speed of light and $\alpha \approx 1/137$ is the fine structure constant). This means that a collision between a proton projectile and a hydrogen target is considered fast if the projectile energy is much above 25 keV (corresponding to $v_i = v_0 = 1$ a. u.). Then energy and momentum changes of the transferred electron during capture can be large compared to those in the bound states (large ΔE in Figure 8.1). As energy and momentum must be conserved in the collision a relatively large amount of energy must be released. Therefore, capture cross-sections are determined at high velocity by very different mechanisms and properties of the bound states compared to those functioning at low collision energy.

At high projectile velocity, one would expect first-order calculations of electron capture to be appropriate, since the target atom wave function should almost not be distorted by the fast projectile. There is very little time available during the collision for the electrons to respond to the motion of the nuclei. This explains the long-standing belief in descriptions of capture by first-order theories, where electron capture is treated by selecting the high electron momentum components in the wave functions of initial and final states for capture. Still in his 1948 paper, N. Bohr considered first-order quantum mechanical treatment to give the correct answer[10]. Figure 8.2 shows a schematic view of the momentum overlap of target and projectile initial and final bound states, respectively.

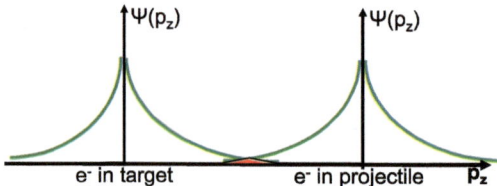

Figure 8.2: Generic view of the momentum overlap of target and projectile initial and final bound states in capture, respectively, that determines basically the cross-sections.

The overlap integral in momentum space essentially determines the capture probability within these first-order theoretical models. This is basically the approach of the well-known Oppenheimer–Brinkmann–Kramers (OBK) approximation [3, 4]. The first-order perturbation theories were developed with modifications in, e. g., the interaction of the active electron with the projectile, the distorted wave approximation [11, 12] and by including higher-order contributions in the eikonal approximation [13]; or by considering the correct asymptotic boundary conditions[14]. Further issues were the use of different interaction potentials before and after collision (the so-called post-prior discrepancy) and the non-orthogonality of the wave functions of the system in the initial and the final channels.

All descriptions have in common that the capture cross-section decreases very rapidly with increasing projectile velocity because of the fast decrease of the electron momentum wave functions with increasing electron momentum. Asymptotically for high v_i in first-order perturbation theories [8] it is for s-states

$$\sigma_{1B} \sim n^{-3} Z_p^5 Z_T^5 / v_i^{12} \sim n^{-3} Z^5 E^{-6}, \tag{8.1}$$

Z_p and Z_T are projectile and target nuclear charge, respectively. Because of the electron momentum wave functions, it is clear that for increasing v_i preferential inner-shell target electrons are captured (n^{-3}). The electron-shell structure of the target atom becomes substantial.

Let us consider as an example the electron capture to protons from Ar target atoms. Figure 8.3 shows the proton energy dependence of the total electron capture cross-section (σ_{CT} crosses) and selectively for capture from the argon K-shell (σ_{CK} circles) [15]. Very obvious is the strong decrease of the total capture cross-section with energy, roughly $\sigma_{CT} \propto E^{-5}$. This is quite well in accord with the Brinkman–Kramers approximation by Nikolaev [16] that adds up theoretical contributions to σ_{CT} from specific shells for the capture of K-, L-, and M-shell electrons of argon (full lines labelled M, L, and K). The cross-section for argon K-shell electron capture makes a small contribution to the total cross-section for capture, as well as to Ar K-shell ionization σ_{VK}, but becomes at higher energies, i.e., above 12 MeV proton energy the dominant capture process. This verifies the importance of the high momenta in the bound electron wave function in the validity regime of first Born approximation. However, there is still another mechanism coming into play at very high projectile energy, where the high momentum components in the wavefunction died out. There a different process (besides radiative electron capture) takes over as we can see below.

For an electron to be transferred in fast collisions, a mechanism was searched and found by Thomas (1927) [17] that accelerates the target electron to the matching velocity of the projectile. He applied a classical kinematic double scattering mechanism (Thomas process). The electron is scattered first by the projectile nucleus and then by the target nucleus in such a manner that the electron finally has almost the same velocity vector as the projectile (as depicted in Fig. 8.4). After this double scattering has occurred, the electron and projectile nucleus can become bound by their mutual attraction. In this mechanism, the target nucleus interacts with the electron but not with the projectile nucleus, and since the electron-nucleus mass ratio is very small, the momentum transferred between the electron and both nuclei is also small. The kinematic conditions for scattering the target electron from the projectile and target nucleus into the direction of the projectile with matching velocity results in a small well-defined deflection of the projectile by an angle θ given by the mass ratio of electron to the projectile.

In the first collision, the target electron (initially quasi at rest) is scattered by the projectile at an angle β to receive the velocity of the ion (see schematic view in Figure 8.4). In the second collision, it scatters at the target nucleus into the direction of the projectile to be captured. With these boundary conditions, one can solve the six equations of constraint for both collisions and determine the particle momenta. The allowed values of v and v_2 depend on the projectile, captured particle, and target masses M_1, m, and M_2, respectively. For example, in the case of the transfer of an electron from atomic hydrogen to a proton, i.e., $p^+ + H \rightarrow H + p^+$, it is easily verified that the angles are $\theta = (m/M_1)\sin\phi$, $\beta = 60°$, and $\varepsilon = 120°$, where $m = m_e$ electron

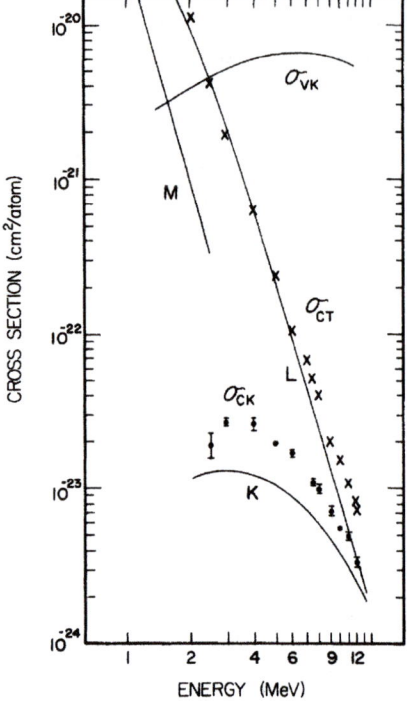

Figure 8.3: Cross-sections for electron capture and Kshell vacancy production in argon by 2.5–12-MeV protons (from [15]). Experimental cross-sections for total capture (σ_{CT}) are shown by crosses and for K-shell capture (σ_{CK}) are shown by full circles with error bars. The curves labeled K, L, and M give theoretical cross-sections for capture of electrons from specific shells [16]. Vacancy-production cross-sections are shown by the smooth curve labeled σ_{VK}.

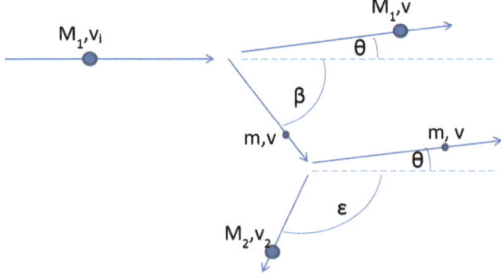

Figure 8.4: Schematic diagram of the classical kinematic double scattering mechanism (Thomas process). It is projectile, captured particle, and target masses M_1, m, and M_2, respectively, with their corresponding velocities v (for details, see text).

mass and $M_1 = M_2 = M_p = 1836m$ (proton mass), $\sin 60° = [(\sqrt{3}) \div 2] = 0.866\theta = (1/1836) \times 0.866 = 4.717 \times 10^{-4}$ rad $= 0.027$ deg. From this kinematical picture, one expects in the differential cross-section of electron capture in high energy p–H collisions a peak at this very small angle. It is the so-called Thomas Peak. In the Ruby Mountains in Nevada, the second highest peak of 3500 m height is named Thomas Peak, to be mentioned here for completeness.

Ever since the early days of quantum mechanics, there was large theoretical interest in this problem (see paper 7 for a number of references). It was, however, not until 1955 when Disko [18] realized that second-order Born is the quantum analogy to the classical Thomas scattering process. Since then, there were numerous different approaches in treating the electron rearrangement at high velocity. These consist of higher-order perturbations, e. g., the second-order Born (B2) [19, 20, 21], where the intermediate electronic states are described by plane waves, and the strong potential Born (SPB) [22, 23, 24], where the intermediate electronic states are described by Coulomb waves that incorporate one of the two potentials to all orders. In SPB-FP (full-peaking (FP) calculation), further peaking approximations to the SPB amplitude are introduced to evaluate the integrals [23]. The impulse approximation (IA) [7, 25] can be deduced from the SPB amplitude by neglect of off-energy contributions. In contrast with the Born-series calculations, the continuum distorted-wave (CDW) [11, 12] approximation describes the electronic wave functions by appropriate distortion operators, and eikonal approximations (EA) [13] is a higher-order calculation with eikonal wave functions. More recently, two different theoretical approaches were added. For proton-helium collisions at 7.4 MeV by Adivi and Bolorizadeh [26], who applied the Fadeev approximation; and by Abufager et al. [27], who based their calculation on the continuum distorted wave approach.

The higher-order calculations show a weaker asymptotic velocity dependence (proportional to v_i^{-11} instead of OBK to v_i^{-12}). Therefore, second order may dominate the nonradiative or so-called mechanical capture cross-section at high energies. In the case of 1s to 1s capture for p on H, this is expected to occur for energies higher than about 50 MeV. We have here one of the few cases in physics where a higher term in the perturbative treatment dominates over the first-order term. The cross-sections are, however, very small at such high energies, which makes the experimental verification of the higher-order dominance in the total cross-section, by an experimental verification of the asymptotic v_i^{-11} velocity dependence versus the first-order Born of v_i^{-12} difficult. The appearance of such an asymptotic velocity dependence in the total cross section seems also to be questionable if one considers, first of all, the contribution from radiative capture that varies with to v_i^{-2} [8], and the necessary relativistic treatment of electron capture at high velocities [28]. A relativistic second-order calculation predicts [28] for this energy range a weaker velocity dependence than nonrelativistic calculations.

An attempt to find experimentally the asymptotic energy behavior was tried by Schwab et al. (see Figure 8.5 and Figure 8.6) for protons capturing from atomic hydrogen [29]. There the asymptotic region is reached at a few 10 MeV energy already. As we see in Figure 8.5, the differences in the slopes between first and second Born approximations is rather small and within the experimental error bars. In order to make

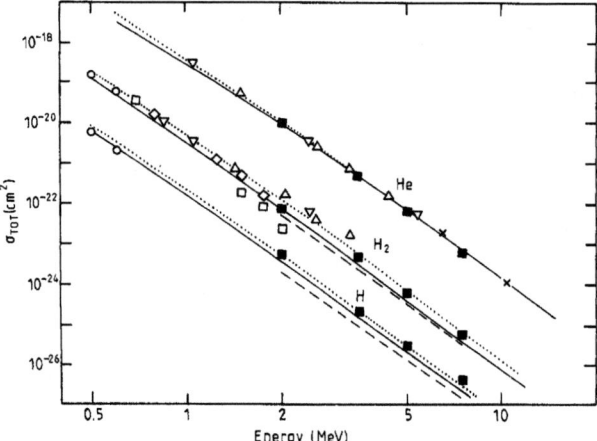

Figure 8.5: Energy dependence of electron capture cross-sections σ_{TOT} for p in H, H_2, and He. Results from [29] full squares, and: open circles [30], triangles up [31], triangles down [32], open squares [33], diamonds [34], crosses [35]. Theoretical results: (-), CDW calculation [12] for capture from 1s into all states; (- - -), SPB calculation by McGuire et al. [24]; (.....), TFBA calculation by Belkic et al., [12] only for 1s to 1s capture. The theoretical results for p on H_2, are obtained from those for p on H by multiplying by 2. In the case of H_2, the cross-sections are multiplied by 10; in the case of He, by 100.

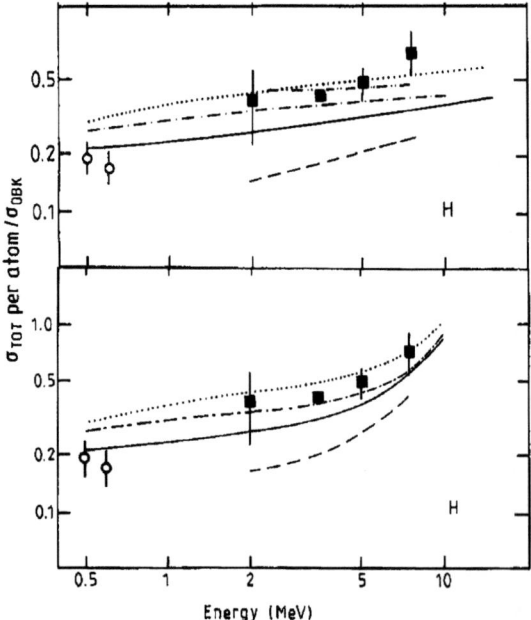

Figure 8.6: Ratios of the cross-sections σ_{TOT} to the OBK cross-sections for p in H (same experimental results [29] and theoretical curves with the same symbols as in Figure 8.5). Additional: (- . . . -) EA calculation by Chan and Eichler (1979) [13]; (- . - . .), SPB-FP calculation by McGuire et al. (1985) [24], both for 1s to 1s capture. Lower plot radiative capture added [8].

a possible deviation from OBK v_i^{-12} scaling better visible in Figure 8.6 the ratios of the cross-sections to the OBK cross-sections are plotted. It is well known that OBK overestimates the capture cross-section by a large factor, therefore, it is not surprising that the ratio is below 1. A small rise of the cross-sections over OBK with energy is seen which could indicate v_i^{-11} instead of the OBK v_i^{-12} scaling. However, the experimental error bars are quite large and there is a substantial contribution by radiative capture at high energies (see lower plot in Figure 8.6). More recently, another attempt was made that demonstrated the v_i^{-11} scaling more clearly (see below).

A more significant evidence for the higher-order contributions to capture can be seen in the angular differential cross-section. The Thomas Peak has been first observed in two different experiments, in proton-helium collisions at 7.4 MeV by Horsdal-Pedersen et al. [36], and in proton-atomic hydrogen collisions at 5 MeV by Vogt et al. [37]. In Figure 8.7, the measured angular-differential capture cross-section $d\sigma/d\theta$ of 5 MeV protons in atomic hydrogen is shown. A clear peak besides the forward peak at around 0.1 mrad appears at an angle of about $\theta = 0.47$ mrad (the Thomas angle). In 5.0-MeV p on Ar, also measured in [37], only the forward peak at around 0.1 mrad is seen. In Ar, the Thomas Peak is covered by OBK-type capture from the much higher Ar 1s state momentum components.

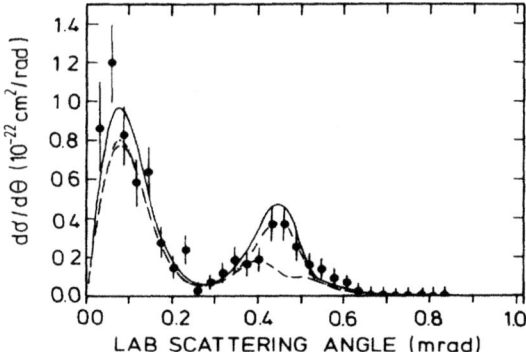

Figure 8.7: The measured [37] angular-differential capture cross-section $d\sigma/d\theta$ of 5 MeV protons in atomic hydrogen. Solid line: SPB-FP calculation [24] for capture from 1s to 1s state. Dot-dashed line impulse approximation for capture from *ls* to *ls* state. Dashed line: CDW calculation [11] for capture from *ls* to final states up to $n = 4$.

Considering the absolute scale, all higher-order theories, folded with the experimental angular resolution, agree well with the data in Figure 8.7. The SPB-FB and IA agrees somewhat better than CDW in the Thomas Peak region, Except CDW that includes non-radiative capture from 1s to final states up to $n = 4$, the others take only nonradiative capture from the 1s target state to the 1s projectile state into account. Not shown here are results of the second Born approximation (B2) which overestimates the cross-sections.

As described above, the Thomas double scattering process in the classical description predicts the correct kinematical angle and the correct velocity scaling. One would expect that a Classical Trajectory Monte Carlo (CTMC) calculation should also describe the differential cross-section well. This was studied in [38] and found that in the classical trajectory calculation the Thomas process is buried under an overestimated OBK-type process. This is due to deep bound states which appear in classical simulations for any collision energy. There the electron has momenta that coincides with the required matching momentum. Because the Thomas process has no preference for large-momentum components, it is not enhanced by the peculiarity of the classical bound states [38] and the Thomas Peak becomes invisible in the differential cross sections calculated with CTMC.

The shape of the differential cross-section around the Thomas angle is somewhat blurred by the experimental resolution. In the comparison to calculations, this was considered by an appropriate convolution of the theoretical data. For the p-He data of [36], however, a stringent test of calculations was problematic. There was thus a strong need for an experiment with higher resolution. More recently, a new and experimentally very different attempt was made to get higher quality results for the helium target. This was done by using COLd Target Recoil Ion Momentum Spectroscopy COLTRIMS [5, 6] and an ion storage ring [39]. The COLTRIMS technique has proven to be a powerful tool to investigate the details of the dynamics of atomic collisions. Unlike in earlier experiments, where the very small projectile scattering angle θ was measured directly, the scattering angle is determined by COLTRIMS from the measured momentum transfer to the recoiling target ion. The excellent luminosity in a storage ring allowed to measure up to high projectile energies of 12.5 MeV with much better resolution and statistics than in the previous experiment. This is shown in Figure 8.8 for 7.5 MeV proton-helium collisions by a comparison of the new data set (triangles [40]) with the data from [36] (circles). The full line is obtained from the newer data by folding with the experimental resolution of [36]. The inset shows the projected recoil momentum distribution on the plane perpendicular to the beam axis with the forward peak in the center and the ring from Thomas scattering.

As clearly seen in Figure 8.9, the newer experimental results from [40] allow a much more detailed comparison to theoretical data than the older dataset from [36]. The red dotted curve is from calculation by Abufager et al. [27] and the full red curve after convolution with the experimental width. The blue dotted curve shows a calculation by Adivi and Bolorizadeh [26] which becomes dashed blue curve when convoluted with the experimental width. Obvious is now the sensitivity to the position of the minimum between forward peak and Thomas Peak, for a comparison with theoretical models (Figure 8.9).

This minimum between the two peaks is not just from competing contributions of first- and second-order cross-sections but from an interference of transition amplitudes [25]. The forward electron capture peak at $\theta = 0$ is thus not associated only

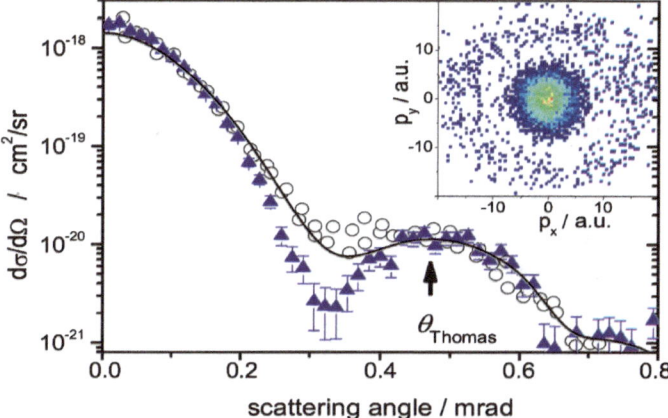

Figure 8.8: Differential cross-sections $d\sigma/d\Omega$ of electron capture in 7.5 MeV proton-helium collisions (triangles, [40]) compared to data from Horsdal-Pedersen et al. [36] (circles). The solid curve represents triangles convoluted with the experimental resolution from [36] and the inset shows the recoil ion momentum distribution in the plane perpendicular to the beam [40].

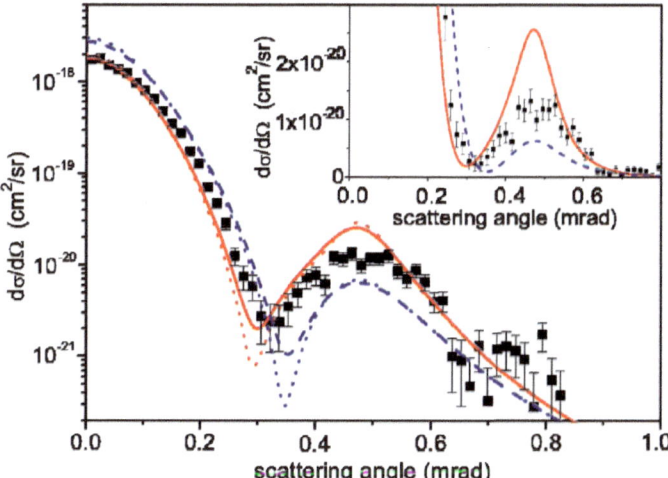

Figure 8.9: Experimental data for 7.5 MeV proton-helium collisions [40] compared to CDW-EIS results from [27] solid curve and the second-order Faddeev calculation from [26] dashed curve. Both theoretical cross-sections were calculated for 7.4 MeV impact energy and were convoluted with the experimental resolution. The unconvoluted calculated data is shown as dotted curves. As inset, the Thomas Peak is shown on a linear scale together with the convoluted theories.

with the OBK mechanism but has also contributions from the second-order amplitude. However, calculations within the first-order approximation do not display any structure in the cross-section around the Thomas angle. The peak at the Thomas angle may, thus, almost exclusively be accounted for by second-order contributions. This fact was used in [41] to check the velocity dependence in these two regions separated by the minimum.

The analysis of the data [41] for this evaluation is shown in Figure 8.10 where the differential cross-sections for p-He nonradiative single-electron capture are plotted for three different energies in the alternative representation $d\sigma/d(\theta^{-1})$ as function of $1/\theta$. Clearly, the forward peak at around $1/\theta = 6$ decreases relative to the Thomas Peak at around $1/\theta = 2$ with increasing beam energy. The position of the minimum, however, does not shift and the cross-sections in the two regions can be integrated separately. These cross-sections [41] are shown in Figure 8.11 multiplied with v_i^{12} to balance the OBK asymptotic velocity scaling. If the forward peak would be purely from first-order capture, it should be approximately constant in the asymptotic velocity region. However, theory does not offer a simple prediction for the velocity dependence in this region. The peak at the Thomas angle is solely accounted for second-order contributions giving a v_i^{-11} asymptotic velocity dependence which is indicated by the dashed line [41].

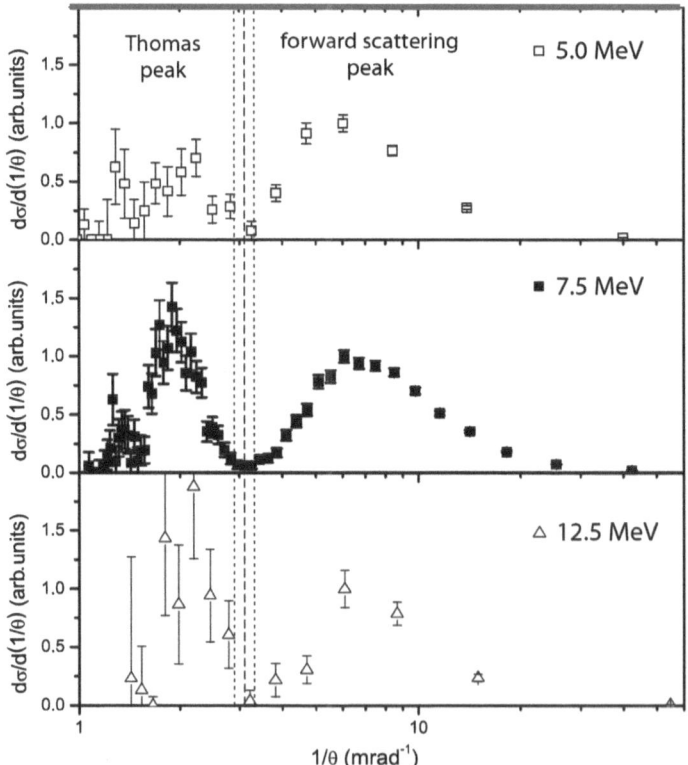

Figure 8.10: Differential cross-sections in the alternative representation $d\sigma/d(\theta^{-1})$ for p-He non-radiative single-electron capture collisions [41]. The vertical lines indicate where the separation between the two peaks is made as well as the error estimate for this separation.

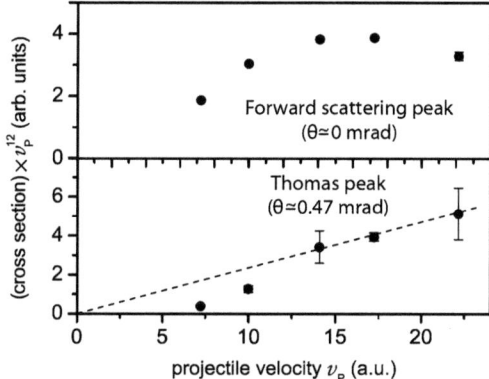

Figure 8.11: Total cross-sections [41] multiplied by v_i^{12} for the nonradiative single-electron capture forward scattering peak (upper panel), and the Thomas single-electron capture peak (lower panel) in p-He collisions (as function of the projectile velocity $v_i = v_p$). The dashed line indicates a v_i^{-11} scaling with the projectile velocity (cf. text).

In electron capture from He or higher atomic charge targets additional capture mechanisms can occur that operate via the interaction with a spectator electron in the atom. One of them is similar to the Auger decay-alternative to the radiative decay of an atomic vacancy. Above we mentioned REC which is in essence due to the ion-charge making the initial state of the electron in the atom unstable with respect to the spontaneous radiative decay into a bound state of the ion. Likewise the Auger process, this decay of the electron in the atom into a bound state of the ion can also occur via the emission of another electron from the atom. The electron to be captured can get rid of its excess energy and reach velocity matching with the ion, when the other electron is emitted backwards. Radiative capture appears already in first-order perturbation theory, and thus, also this electron-electron interaction mediated capture is a first-order transfer mechanism [42].

In case of a He (or higher Z_A) target there is also an additional higher-order mechanism due to the interaction with another bound electron. The primary electron can scatter off this other electron in, e. g., He instead of the target nucleus and get the momentum to match the projectile velocity for getting captured. The change of the transverse momentum will be somewhat smaller than in the original Thomas peak, so the projectile is scattered at a smaller angle. Evidences for this higher-order process were found, e. g., in [43, 44, 45].

In Figure 8.10, one sees below the Thomas Peak region consistently a small peak (also at lower beam energies, as shown in [46]). This and the observation that the angular position of the forward peak is independent of the projectile energy lead the authors of [46] to the following analysis: A synthetic capture probability P as function of the scattering impact parameter b was adjusted to the forward peak. That $P(b)$ showed a decreasing P and falls off with b faster for increasing beam energy, as one would expect from such a capture process. Therefore, the capture cross-section can be viewed as the projectile has to pass through an aperture around the atom in order to capture with a certain probability an electron. In [46] they replaced the synthetic $P(b)$ by a step function $P(b)$ with constant P_0 and a cut-off impact parameter b_{max}. These were also adjusted to match the forward peak.

In analogy to light diffraction in a circular aperture, the diffraction pattern was derived [46] by a Fourier transformation of the $P(b)$ step function. This is plotted together with the measured $d\sigma/d\Omega$ in Figure 8.12 for the 7,5 MeV P – He data [46]. The measured $d\sigma/d\Omega$ has obviously similarities with the diffraction pattern of a monochromatic beam with de Broglie wavelength of 7.5 MeV p that passes through an aperture with b_{max} radius. A zero's-order maximum (the forward peak) then the first-order maximum at around 0.4 mrad, covered partly by the Thomas Peak, and a second-order maximum at 0.75 mrad which is very close to the third unexplained peak in the data. In the next section, we shall see that these diffraction effects can also be observed much more pronounced in a double slit experiment, where the atom as a single scattering center is replaced by a two-atomic molecule.

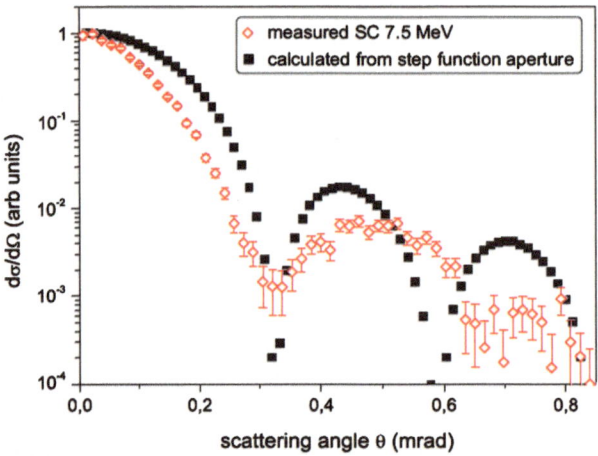

Figure 8.12: Measured cross-section $d\sigma/d\Omega$ (open diamonds) and cross-section resulting [46] from a step-function $P(b)$ (full squares) with radius chosen to give a minimum in $d\sigma/d\Omega$ at 0.32 mrad.

3 Interference effects in electron capture from molecules at high velocity

Double-slit experiments been performed with neutrons [47], metastable helium atoms [48], and the wave nature of the propagation of still heavier species of matter such as fullerenes and even larger molecules [49] was demonstrated. In ion-atom collisions, two-path-ways (in which different "reaction trajectories" may lead to the same final state) give rise to, e. g., Stueckelberg oscillations (see, e. g., [50, 51]) and impact parameter dependent interference effects [52] and see Chapter 4. Recently, two center interferences were also observed in electron emission in ion-H_2 collision experiments [53, 54]. The de Broglie wavelength for MeV energy ions is orders of magnitude smaller, therefore, demonstrating a Young-type interference for such high energetic beams requires a double slit as narrow as the internuclear distance in molecules. In fast ion-atom collisions the de Broglie wavelength of the hydrogen atoms formed by capture, as discussed in Chapter 2, is only of the order of 25 fm. The capture "occurs" in the vicinities of 0.3 a. u. (1 a.u. is $a_0 \approx 5.3 \times 10^{-10}$ m the Bohr radius) from the nucleus (Chapter 2 and [55]). The nuclei in H_2 are separated by $a = 1.4$ a. u. Thus, the superposition of the two contributions to the outgoing de Broglie wave can lead to interference effects in the intensity of the formed H atoms dependent on the molecular orientation.

Tuan and Gerjuoy [56] calculated the capture cross-section from a hydrogen molecule by a proton, and found that the cross-section is somewhat larger than that of the sum of two hydrogen atoms. They calculated the molecular capture cross-section as a coherent sum of two transfer amplitudes centered on each atomic center, but integrated over all molecular orientations and missed a dependence of the cross-section on the orientation of the molecule. Later, several authors [57, 58, 59] published results showing that the capture cross-section should depend on the orientation of the molecule. This dependence would be a consequence of a quantum mechanical interference of amplitudes of capture of an electron localized at one of the two atoms of the molecule. The concept and descriptions of the molecular states and initial localization of the captured electron is quite approximate in these calculations, thus it was not self-evident that an interference pattern is visible in experiments.

One has to measure the capture cross-section as function of the molecular orientation relative to the projectile direction. As the molecule gets charged and excited, it can dissociate and there is the possibility to derive its orientation from analyzing the fragments. When the H_2^+ molecular ion is formed in the $2p\sigma_u$ excited state it will rapidly dissociate into a proton and a ground state hydrogen atom ($X^{q+} + H_2 \rightarrow X^{(q-1)+} + H_2^+ \rightarrow X^{(q-1)+} + H + p$). The dissociation is along the internuclear axis and it is well known that each fragment gets roughly 8 eV kinetic energy. The experiment has to measure the fast charge changed ions to select the capture and from the same collision the full velocity vector of the proton from dissociation of the molecule. With a COLTRIM spectrometer mentioned above, this is possible [60]. The velocity vector of the proton gives the molecule orientation, as the capture is very fast (10^{-18} s) and also the dissociation process is faster (10^{-15} s) than the rotation time (10^{-12} s) of the molecule.

A first experiment was done at an accelerator with bare oxygen (O^{8+}) ions [61] for exploiting the Z^5 increase in the cross-section compared to proton projectiles. An angular dependence of the cross-section was indeed observed in qualitative agreement with theory. They observed the maximum cross-section when the molecule was oriented perpendicular to the beam axis. In a calculation [62], better agreement was found with a model that also takes excitation into account compared to a model with only single capture. This seems reasonable, considering the high projectile charge and consequently high perturbation of the molecule. A much better test of the calculations should be to measure capture from molecules by protons and also to detect the diffraction in the formed H atoms. For that, a higher luminosity is needed which can be provided by a storage ring. The experiment was successfully performed at the CRYRING storage ring [60]. Several experimental challenges had to be mastered, such as the open COLTRIMS for injection of the beam in the storage ring, the high rate of H_2^+ ions by ionization, and collecting and analyzing the fragments with the large dissociation energy. Figure 8.13 shows a schematic view of the COLTRIMS set up in CRYRING [60].

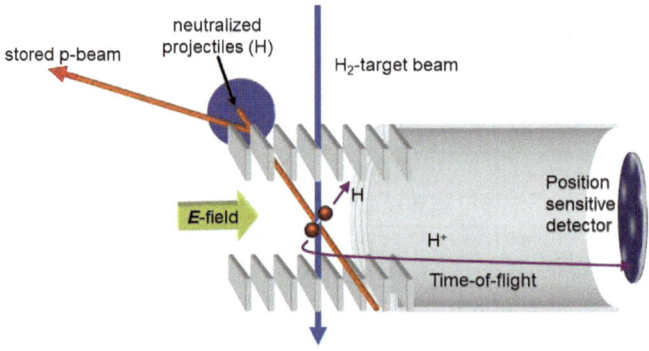

Figure 8.13: Schematic view of the COLTRIMS set up in CRYRING [60].

In the upper panel of Figure 8.14, we see the time-of-flight spectrum [63] for 1.3 MeV p on H_2 molecules, and it is clear that the proton peak is much broader than the H_2^+ peak. This is because of the high kinetic energy release of the dissociation process. In the lower panel of Figure 8.14, the time-of-flight (TOF) information is combined with position information from the position sensitive detector (PSD, see Figure 8.13). We see (Fig. 8.14) that the protons form a ring when combining the displacement on PSD with TOF. In its center is a proton peak from proton shake-off after capture. Its intensity is originally much higher and has been blocked by a screen in this area of the detector [60]; also, the intensity of the narrow H_2^+ peak at 575 ns TOF which is mostly from ionization random coincidences was blocked. Obviously, from the time-of-flight measurement, one cannot distinguish the origins of the target protons, but together with the positions one gets all three velocity components. And, from the length of the velocity vector one distinguishes between fast protons from the $2p\sigma_u$ dissociation and slow protons from ground state dissociation.

Thus one can orient the dissociating molecule with respect to the beam axis. For the further analysis of the capture cross-section dependent on the orientation angles, we need to introduce a coordinate system at the center of the H_2 molecule.

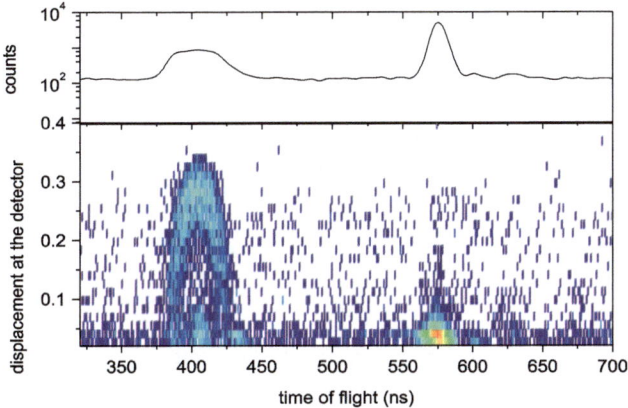

Figure 8.14: Upper panel: the time-of-flight spectrum for 1.3 MeV p on H_2, lower panel the time-of-flight combined with position information from the position sensitive detector (see Figure 8.13) [63].

The projectile moves here in the xz-plane at a distance b (impact parameter) from the molecule center along the z-axis. Detecting the proton velocity vector, one gets the orientation (ϑ, ϕ) of the H_2 molecule at the time of the collision and one detects the corresponding fast H atoms on the neutralized-projectile position-sensitive detector (see Figure 8.13). Fixing the angle ϕ of the molecule defines also the orientation of a (x', y') coordinate system on this detector. When, e. g., $\phi = 0$ is selected, one expects interference patterns in the probability distributions for fast H-atoms along the x'-axis which changes for different values of ϑ. For $\vartheta = 90°$, the phase shift on the z-axis is zero and the partial waves interfere constructively, there should be a sharp peak at $x' = 0$ on the neutralized-projectile position-sensitive detector (see Figure 8.13), with small side bands along x'. The resolution of the experiment [63] was not good enough to resolve the interference structure in the fast H atom beam patterns on the detector. This would have been the analog of the light diffraction pattern on the screen in an optical double-slit experiment. However, the intensity distributions of fast H atoms along the x'-axis could be fitted and a width, called in [63] "peak sharpness" could be extracted.

These experimental results are plotted in Figure 8.16 as open triangles (peak sharpness), and as full squares (orientation-angular-differential cross-section). One clearly sees a variation of the experimental capture cross-section with the orientation of the molecule, and also a corresponding variation of peak sharpness. This strongly indicates an interference of the capture amplitudes from the two atoms of the molecule, In order to understand the phase shift that leads to the observed pattern, one needs to consider the transfer of forward momentum to the projectile due to the electron capture. At the velocity v_i, the projectile gains forward momentum p:

$$\delta p = nmv/2 + Q/v_i \tag{8.2}$$

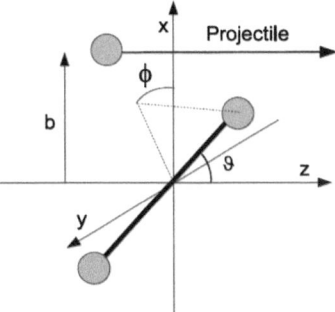

Figure 8.15: Coordinate system at the center of the H_2 molecule for definition of its orientation with respect to the projectile axis that coincides with the z-axis.

where n is the number of captured electrons, Q is the inelasticity in the considered reaction, and with the given v_i, this term can be neglected. The change in wave number k is from (8.2): $\delta k \approx v_i/2 = 3.6$ a. u. (this is a very small change of the projectile wave number of $k = 1.3 \times 10^4$ a. u.). With that, one gets a total phase shift $\delta \Phi$ on the z-axis between the two capture amplitudes at the two atoms of the molecule composed of the change in projectile momentum, or wave number δk, and the distance of propagation between the target protons:

$$\delta \Phi = \delta k \cdot a \cdot \cos \vartheta \approx 5 \cos \vartheta, \tag{8.3}$$

and for $\vartheta = 51°$ and $129°$ these phase shifts are π and $-\pi$, respectively. One expects thus destructive interference at these angles, which is close to what is found in Figure 8.16 (from [63]). There the upper horizontal axis gives this estimated phase shift. Possible small deviations of the minima from the π and $-\pi$ positions may be due to contributions from the neglected Q/v_i term in (8.2) and that the effective "slit" distance' may be smaller than the nominal $a = 1.4$ a. u. value due to an enhanced electron density with higher electron momentum between the H_2 nuclei.

The full line in Figure 8.16 is from the theoretical prediction [59] for single-electron capture from the molecule. The experimental results agree well with that theory which only considers capture. As the experiment exploits also the excitation process for orienting the molecule, one may conclude that the probability for excitation of the remaining H_2^+ ion does not depend strongly on ϑ. One should also note that the minima in the capture and excitation cross-section are not found for molecules aligned with the projectile (z-) axis, instead at 51 and 129 degrees, thus steric (geometrical shadowing) effects can therefore not be the reason for the observed oscillations.

An interesting question arises: Which is the result, when the projectile is forced to pass through both slits always. Or, in the concept of these experiments, which is the pattern in the cross-section variation with the molecular axis orientation for double capture from H_2, assuming that capture occurs at different well-defined positions in

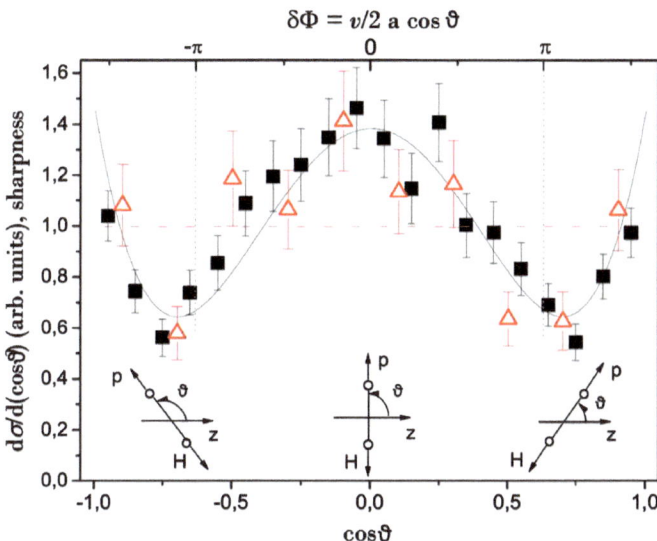

Figure 8.16: The orientation-angular differential cross-section (from [63]) for capture and excitation in 1.3×1.3 MeV p on H_2 (measured solid squares), and projectile peak sharpness (open triangles, cf. text) as functions of cosine of orientation angle. The upper horizontal axis gives the expected phase difference ($v = v_i$) for zero angle projectile scattering. The full curve is the theoretical result of [59] for single-electron capture.

the molecule? This case of $He^{2+} + H_2 \rightarrow He + H_2^{2+}$ was studied by Martinez et al. [64] and Misra et al. [65]. In [64], at low impact energy no indication of interference was found, whereas in [65] the impact energies of around 1 MeV with the storage ring CRYRING were measured and interesting interference patterns were found. The result of double capture from the H_2 target molecule is instantaneous dissociation in two protons with 9 eV dissociation energy each. Compared to the dissociation time scale, the collision time is two orders of magnitude shorter and, therefore is the dissociation symmetric with respect to the molecular axis. It is thus enough to detect only one of the protons for fixing the internuclear axis of H_2. In the experiment of [65] a little more than half of the detector was covered by a foil to reduce random background from ionization and also to avoid two protons hitting the detector at the same time which destroys position information in the signal. Therefore, the data points for positive and negative $\cos \vartheta$ are mirrored in Figure 8.17 which shows the measured orientation-angular differential cross-section for double capture of $He^{2+} + H_2$ at two different beam energies [65].

The rather monotoneous behavior of the double electron capture cross-section at 300 keV (filled triangles) changes dramatically when going to 500 keV/u where the cross-section goes through a strong sinusoidal oscillation (open squares). The experimental results are fitted with the function (dotted curves in Figure 8.17)

$$d\sigma/d(\cos \vartheta) = A[1 + V \cos(g \cdot \delta\Phi)], \tag{8.4}$$

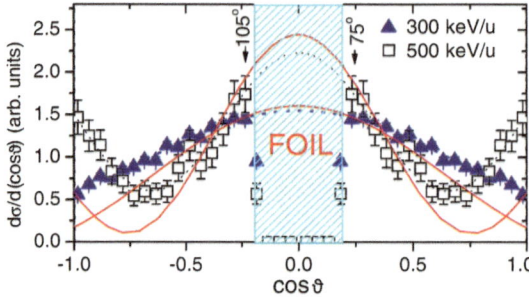

Figure 8.17: The measured orientation-angular-differential cross-section (from [65]) for double capture of $He^{2+} + H_2$ at two different beam energies. Dotted lines are fits to the experimental data
using equation (8.4), solid curves are the OBK results (from [65]) which should be multiplied by
8.4×10^{-21} cm^2 and 4.8×10^{-23} cm^2 for 300 and 500 keV/u for absolute scales, respectively.

where A, V and g are free parameters. Here, one makes the assumption that both electrons are captured when the projectile is in the vicinity of either H_2-target nucleus. If
one uses $n = 2$ in equation (8.2), then is the change of the projectile wave number
$\delta k = v_i$ on the z-axis. Using this and equation (8.3) in equation (8.4), then one gets a
reasonable good fit of the experimental results (dotted lines Figure 8.17).

The solid curves in Figure 8.17 are fully quantum mechanical, OBK type calculations [65] for double electron capture from H_2 by He^{2+} ions. They are based on a simple
model incorporating a linear combination of products of hydrogenic 1s wave functions
to describe the two-electron two-center state of the H_2 molecule. The transfer of the
two electrons can be understood, in the OBK approach, as following [65]: one electron is captured as a result of the direct interaction with the projectile close to one
of the nuclei, while the transfer of the other electron becomes possible through the
nonzero overlap between the initial target and final projectile states in a "shake-over"
process, and thus this other electron does not need to be from a region close to the
target nucleus. Obviously, there is a large similarity between double-electron capture,
and single-electron capture processes.

In order to make this more clear, the data is plotted in Figure 8.18 for single-
electron capture (Figure 8.18(a–d) from [60]) and compared with those for double-
electron capture (Figure 8.18(e, f) from [65]). The curves are fits according to equation
(8.4) and for $g = 1$ the vertical dotted lines indicate $\delta\Phi = \pm\pi$ where the first minimum
for destructive interference is expected along the z-axis. In atomic collision physics
one uses a perturbation strengths $S_p = q/v_i$ which is given in Figure 8.18(a–f).

As one sees from Figure 8.18, the deviation from the expected value $\delta\Phi = \pm\pi$
increases with increasing values of S_p. According to equations (8.3) and (8.4), this
could mean a decreasing effective slit separation for increasing values of S_p. Clearly,
the probability density of the target electron cloud is higher at a given distance from
one nucleus and between the two nuclei than at the same distance from the same nucleus in the other direction. This presumably also means that the two-electron transfer

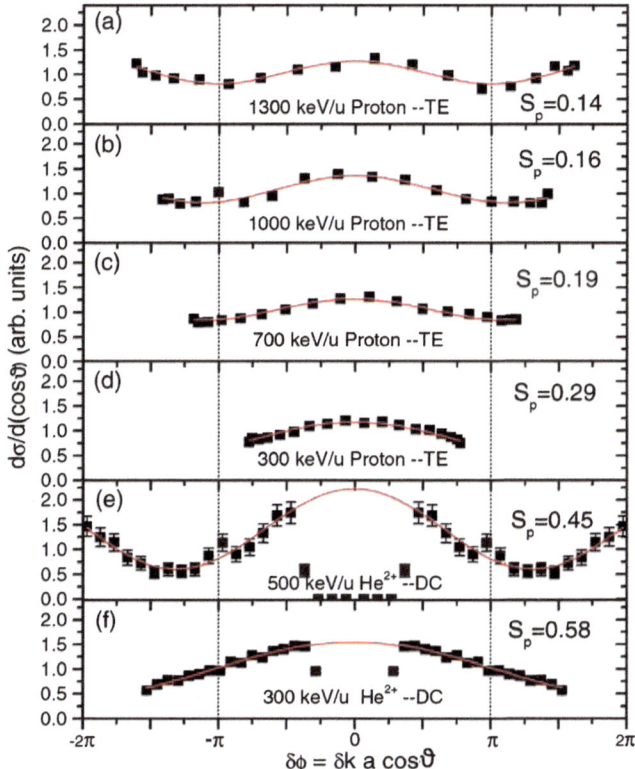

Figure 8.18: Panels (a–d) experimental results for 1300 keV/u–300 keV/u protons on H_2 – single capture, from [60]. Panels (e,f) double capture for 500 and 300 keV/u He_2^+ on H_2 [65]. Curves are fitted functions (equation (8.1)) to the experimental data (filled squares). Vertical dotted lines at $\pm\pi$ show the expected positions for the first minima according to equation (8.4) with $c = 1$.

probability is asymmetric with respect to either target nucleus and that it is higher between them. In principle, this implies an effective slit (or scattering center) separation smaller than a. The larger the two localized regions within which such simultaneous two-electron transfer processes are effective (i. e., the larger S_p is), the larger the deviation from a slit distance equal to the internuclear separation for H_2 becomes.

These interference phenomena with simple molecules were also studied at somewhat lower energies (75 keV or 1.6 a. u.) by Lamichhane et al. [66]. They measured in [66] the differential cross-sections for electron capture in p-H_2 collisions with dissociation via the vibrational excitation channel. Particular focus was set on different projectile coherence lengths and they identified two types of interference, the single center interference and molecular two-center interference. We can understand the single center interference as a diffraction effect at the scattering centers, analog to the one explained above for p-He collisions as in [46]. It appears here only for transversal coherent projectiles as well does the molecular two-center interference. This data on the molecular two-center interference reveals several effects that require further clarifications and go beyond the scope of this review.

L Schmidt et al. [67] studied reactions that are in some sense inverse to those discussed before: $H_2 (1s\sigma)^+ + He(1s^2) \rightarrow H_2^* + He^+ \rightarrow H + H + He^+$. Besides the fact that the molecule is now the projectile (in [67] it had velocity $v_i = 0.45$ a. u.) instead of the target, the direction of the electron transfer is inversed. Nevertheless, an interference can be expected and it turned out to be much richer than anticipated, since electronic degrees of freedom strongly affect the interference [67]. To translate this experiment into the double slit picture, we transform into the moving projectile frame where the He atom can be scattered on both projectile atoms. Here, we can apply the terminology, as above for the orientation of the molecule, to define the phase shift between the two capture amplitudes.

In these experiments, however, the electronic and nuclear degrees of freedom are stronger coupled than in the previous ones we discussed. Therefore, we need to look into the dissociation pathways and electronic degrees of freedom before coming to the capture amplitude interferences. As explained above, the electron capture occurs much faster than the dissociation, so "suddenly" several electronic states of the neutral molecule are being populated and the energy Q is taken from the kinetic energy between He^+ and the H_2 molecule. This Q-value can be measured by the recoil ion momentum of the He^+ ion in direction of the beam axis.

The populated excited states can radiatively decay into a repulsive state of the neutral H_2 molecule, from there it dissociates and converts the rest of its energy above ground state into kinetic energy release (KER). This can be measured by the neutral fragments of the molecule and the KER distribution versus the Q value give information about the dissociation pathways. Without going into details here (which can be found in [67]), the experimentalists selected events with Q+KER of (-15 eV < Q+KER < -10 eV) and with KER between 1 and 5 eV, where they picked only the direct electron transfer from $He(1s^2)$ into the repulsive $2p\sigma$ orbital of the hydrogen molecule that ends into $H(1s) + H(1s)$.

Having well-defined electronic degrees of freedom, it is possible to see the double slit interference in the capture cross-section, but one better switches again now into the frame of the H_2^+ projectile. Now the roles are exchanged, the molecular ion does not capture, but the He atom, as projectile, transfers an electron to the molecular ion. The wave of the He atom with de Broglie wavelength of 0.0019 a. u. scatters at the two atoms of H_2^+ separated by a, and leaves an electron there. The expected scattering pattern consists of interference structures of two spherical waves originating from the two scattering centers. By selecting the KER, the authors of [67] could vary the internuclear distances, and additionally change the orientation of the molecule with respect to the beam axis. Such scattering patterns [67] of transversal momenta $P_{y',He}$ and $P_{x',He}$ instead of scattering angles θ are shown in Figure 8.19. In Figures 8.19(a) and 8.19(c), and the molecule is oriented 90° to the beam axis, while in Figure 8.19(b) an angle of $\vartheta = 55°$ to the beam axis is selected with the same KER as in 8.19(a).

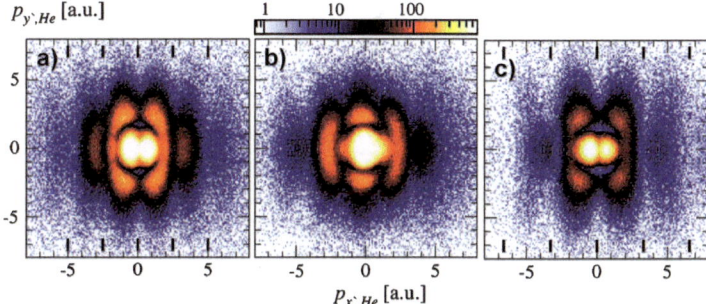

$P_{y',He}$ [a.u.]

$P_{x',He}$ [a.u.]

Figure 8.19: (From [67].) Transversal momenta $P_{y',He}$ and $P_{x',He}$ of the He$^+$ scattering pattern from the reaction 10 keV H$_2$ (1sσ)$^+$ + He(1s^2) $\rightarrow$ H(1s) + H(1s) + He(1s)$^+$ in the plane perpendicular to the beam direction (z-axis). From the relative momenta of the molecular fragments, events, with different internuclear distances and different angles ϑ between the molecular axis and the beam direction from the data, were selected: (a) $80° < \vartheta < 90°$, 1 eV < KER < 2 eV. This corresponds to 2.9 a. u. > R > 2.3 a. u. (b) $50° < \vartheta < 60°$ and 1 eV < KER < 2 eV. (c) $80° < \vartheta < 90°$ and 3 eV < KER < 4 eV (corresponds to 2 a. u. > a > 1.7 a. u.). The thick vertical lines show the positions where interference maxima are expected for the coherent scattering at the two molecular centers. The experimental data [67] have been mirrored at the horizontal axis in order to reduce the statistical error.

For a quantitative comparison between the experimental data and a simple model, the authors of [67] extracted events with fixed ϑ, KER, and $P_{y'x',He}$ to get one-dimensional distributions over the angle ϕ (see Figure 8.15) shown in Figure 8.20 as blue dots. The three cuts from [67] displayed here, correspond to the standard double slit geometry, where the molecular internuclear axis is oriented perpendicular to the incoming beam ($\vartheta = 90°$), and selecting three different choices of a and $P_{y'x',He}$. The fringe separation scales with internuclear distance as expected from an optical double slit. Model calculations of a coherent superposition of two partial waves are made in reference similar to those explained above for the data of [67]. These results are plotted in Figure 8.20 as full and dotted lines where the difference is a possible extra phase shift, i. e., however, very small at this angle of $\vartheta = 85°$–$90°$. The excellent agreement demonstrates again the Young-type interference in collisions of atoms with fixed-in-space molecules; here with the additional variation of the interference pattern with the molecular parameters.

This same group used the previously described experiment, to detect simultaneously the momentum transferred to a "free-floating molecular double slit" and the momentum change of the atom scattering from it, to study [68] a longstanding puzzle in quantum physics. Can the nonlocality that leads to the interference of matter waves that pass through a double slit assembly be removed, as proposed by Einstein [69]: It should be possible to determine the pathway of each individual particle passing through a double slit by observing the recoil momentum it imparts onto a first slit used to diffract the particle wave, still ensuring it coherently illuminates the double-slit assembly. In the experiment of [68], the authors used HD$^+$ molecules instead of H$_2^+$

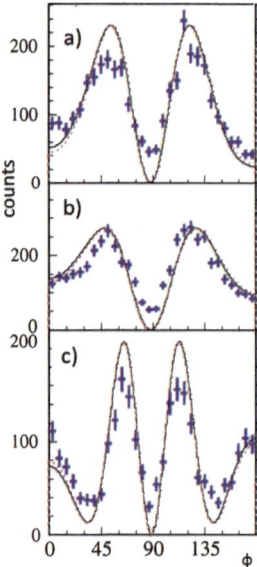

counts

Figure 8.20: (From [67].) Counts of He$^+$ (blue dots), for $\vartheta = 85°-90°$ and for selected transversal momenta $P_{y'x',\mathrm{He}}$, and KER given below, in dependence of the angle ϕ. The solid line is calculated using the full theory (from [67]) and the dotted red line is from a simpler model (from [67]). In each panel, the theoretical curves are normalized to the respective data. (a) 2.2 a. u. $< P_{y'x',\mathrm{He}} <$ 2.4 a. u. and 1.5 eV $<$ KER $<$ 2 eV; (b) 2.2 a. u. $< P_{y'x',\mathrm{He}} <$ 2.4 a. u. and 2 eV $<$ KER $<$ 3 eV; (c) 3.5 a. u. $< P_{y'x',\mathrm{He}} <$ 3.7 a. u., and 2 eV $<$ KER $<$ 3 eV.

to number the slits. They observed Young-type interferences behind the free-floating labeled molecular double slit and measured the momentum transfer that Einstein requested and obtained a good agreement with a classical description of the slits. That description, however, assumed on a microscopic level, that momentum transfer was not ascribed to a specific pathway but shared coherently and simultaneously between both. So in accord with Bohr's view, a quantum mechanical description of the molecular slit dynamics is necessary to describe the observed interference phenomena. As also already described above, [68] showed clearly that momentum transfer from the projectile to the slit changes the interference features in full agreement with predictions from quantum modeling the kicked-molecule slit dynamics.

4 Slow collisions and possibilities of interference in state selective capture

In Figure 8.1, we showed a simplified energy balance for capture, now we consider the case where T, the kinetic energy release in capture, is small compared to the binding energies of the active electron. Then the major parameter that determines the cross-section is the collision trajectory with the distance of approach where the wave functions overlap. In this most simplified approach, one has to find the internuclear distance R_c where the potential barrier formed by the overlap of the two Coulomb potentials decreases to below the binding energy of the electron (see schematic in Figure 8.21). In the case where there are bound states in the ion that have a binding energy close to that of the electron in the atom, it can transfer near resonant. This gives the critical distance R_c, and thus the cross-section $\sigma = \pi R_c^2$. In this case of resonant transfer, the electron capture cross-section can be large, even for producing inner-shell vacancies at low velocities and dominate over direct ionization and excitation, so these channels are not in strong competition with capture.

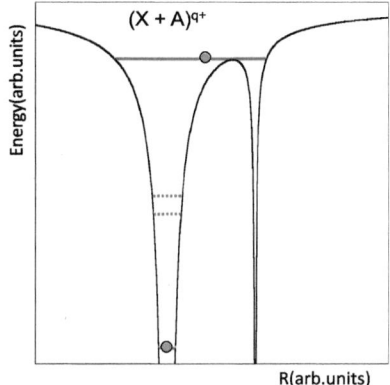

Figure 8.21: Schematic view of the potential barrier that determines the distance of approach for capture in the collision of ion X with target atom A and thus the cross-section for electron capture.

In this part of the article, we shall concentrate on vacancy production in inner shells of atoms by capture into an inner-shell state of the projectile, at low v_i compared to v_B ($v_i \ll v_B$). This process can be resonant in energy (see Figure 8.1) if T is small compared to the binding energies $E_A(n, l)$ and $E(n_1 l_1)$ and $E_A(n, l) \approx E(n_1 l_1)$, e. g., transfer between equivalent states if the ion nuclear charge is about equal to the target atom nuclear charge.

Quantitative values for contributions of inner-shell capture have mostly been derived from taking the rise in the target vacancy production cross-section when the resonant projectile shell is carrying a vacancy. In Figure 8.22, the cross-section for Ar K x-ray emission is plotted as function of the charge of sulphur projectile in collisions of 20 MeV S^{q+} on Ar [70]. In this experiment, an ion velocity of around 5 a. u. is chosen, which is around 1/4 of the Ar K Bohr velocity. Inner-shell electron capture is observed by the sharp rise of target K x-ray emission (here by two orders of magnitude) when the projectile gets highly charged ($q > 14^+$) so it brings K vacancies into the collision.

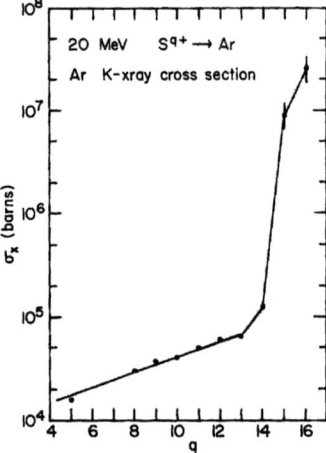

Figure 8.22: K-X-ray production cross-section for the Ar target atom, bombarded with 20 MeV S^{q+} ions as functions of the ion charge state q [70].

A small rise is seen already at charge state 14^+ which is due to a fraction of the projectiles in a 1s2s metastable state.

The weaker rise of the Ar K-x-ray cross-section below $q = 14$ stems from excitation into S L-shell states. This is explained within a model postulating the formation of molecular orbitals during the collision and by $2p\pi\sigma$–$2p\sigma$ rotational coupling between them [71, 72, 73].

In the beginning of laboratory experiments on atomic collision processes, Ziemba et al. [50] found surprising phenomena in K-shell capture differential cross-section. They studied the capture with light ions, such as H^+, He^+, when scattered at fixed angle from light target atoms, such as H_2, He, and varied the ion beam energy in the range of a few keV. The incident ion, after undergoing a single collision which is hard enough to result in a 5° deflection, was analyzed to determine whether it had captured or lost electrons.

In these experiments, Ziemba and Everhart found [50] a rather striking behavior of this capture probability that is shown in Figure 8.23 for He^+ on He, The electron capture probability P_0 plotted vs. incident ion energy reveals seven pronounced peaks. The electron loss probability P_2 also shows peaks alternating with P_0. In a paper by Ziemba and Russek [74], this effect was attributed to resonant capture from the target He K-shell to the projectile He^+ K-shell. These results led in the following years to new concepts of reactions in atomic collision dynamics. The concept of quasi-molecular orbitals was developed [71] to explain the phenomena that were found in slow ion-atom collisions. This development culminated in the predictions [75], proposals [76], and indications [77] of the formation of super-heavy quasi-molecules.

In the theoretical discussion of Ziemba and Russek [74], the nuclei are assumed to move on classical orbits and quantum theory is applied to determine the probability

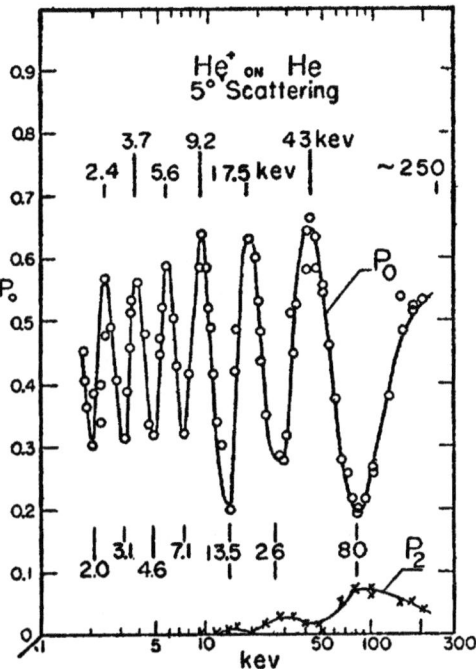

Figure 8.23: The electron capture probability P_0 and electron loss probability P_2 vs. incident ion energy for 5° scattering of He$^+$ on He [50].

of electron exchange as a function of the impact parameter and relative velocity. Since the nuclear charges are equal, the ground-state eigenfunction of the quasi-molecule formed during the collision must be described by a linear combination of symmetric and antisymmetric eigenfunctions. For the case of capture without excitation, the coefficients in this linear combination are constant during the collision, but the relative phase, which depends upon the difference in eigenenergies of the symmetric and antisymmetric states, is altered by the collision. This phase change corresponds to an electron being transferred periodically from one nucleus to the other during the collision. Both the velocity (energy) and impact parameter (scattering angle) variations of the calculated exchange probability were found in qualitative agreement with the data in Figure 8.23. The validity of this approach was also shown by the finding [50] that the interference peaks are nearly evenly spaced in collision time.

A simplified view (from [50]) of the physical situation in these experiments is shown for p on He in Figure 8.24. Many other ion-atom combinations were examined in [50] and found that they exhibit similar phenomena. The angle of 5° was arbitrarily chosen to insure deep interpenetration of the colliding atoms during the collision. The data are not very sensitive to the choice of this angle. It is the time difference between equivalent points of the evolution of the quasimolecule that is formed in the collision (indicated by circles in Figure 8.24).

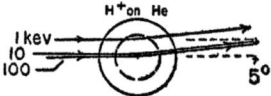

Figure 8.24: Paths of the incident ion at 1, 10, and 100 keV are shown schematically for the H⁺ on He collision (from [50]). The circles indicate radii of the interaction.

After these first experiments on inner-shell electron capture, it took one to two decades until progress toward heavier projectiles was made. This was due to a lack of slow highly charged ions with initial K vacancies and the experimental techniques to study K capture differentially because of its extremely forward-peaked angular distribution. The first impact parameter dependent K capture probability with S^{15+} ions from Ar atoms [78] was still of low quality. Later an experiment with F^{8+} ions on Ne targets was performed that detected K capture by the Ne K Auger electrons [79]. The thus derived impact-parameter dependence for the K-capture probability $P(b)$ and also the Ne K excitation probability when the ion was without K vacancies is plotted in Figure 8.25. The full curve in Figure 8.25 is from a two-state atomic expansion (TSAE) method [80] similar to that by Ziemba and Russek [74] described above for the case of He⁺ on He.

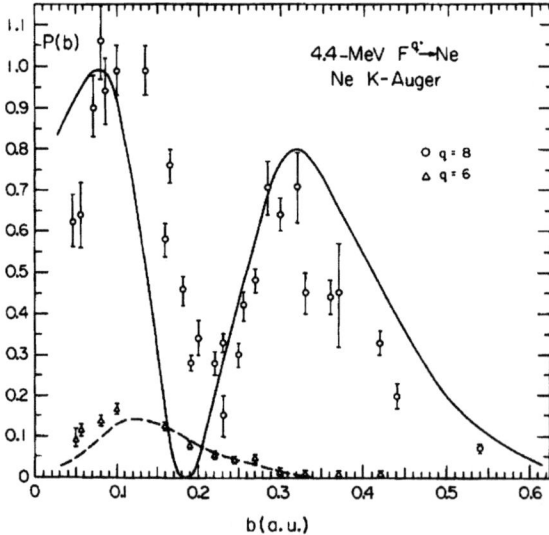

Figure 8.25: Impact-parameter dependence for the Ne K-Auger production probability [77] for projectiles without (triangles, $q = 6$) and with K vacancies (circles) at 4.4 MeV. The broken line represents the molecular excitation probability [72] and the full line is the TSAE calculation from [80].

In Figure 8.22, we saw the steep rise of K-X-ray production cross-section for Ar target atoms when the S^{q+} projectile became hydrogen-like which was taken as evidence for K-shell to K-shell charge transfer. In Figure 8.26, we can see now the variation of the

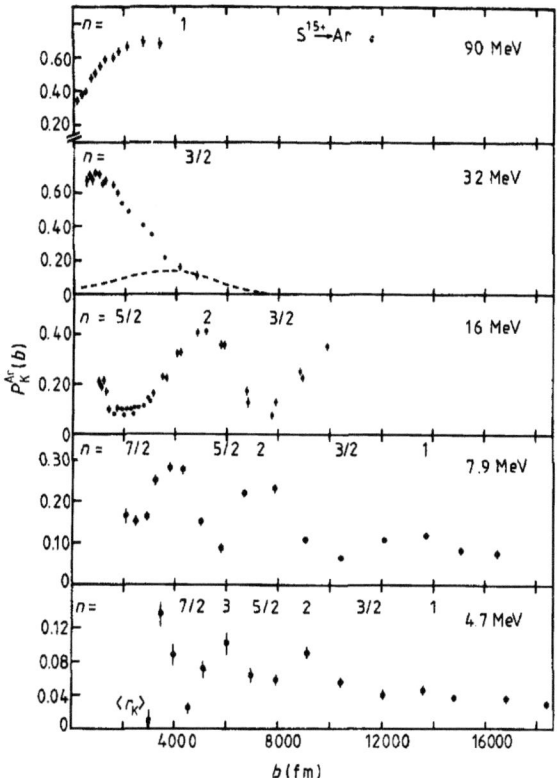

Figure 8.26: Ar K vacancy probabilities P_K^{Ar} as a function of impact parameter b at five different collision energies [52]. The broken curve shows the $P(b)$ from S projectiles at lower charge states [73]. The numbers at the maxima and minima indicate the order of the interference (see text).

corresponding K-capture probability with the impact parameter for different ion beam energies. One can see distinct oscillations of the probability also for this slightly more asymmetric collision system. The damping of the transfer probability can clearly be seen in the data. A probability of near unity as found in the resonant K K transfer collision systems such as He$^+$ on He, or still in the near resonant F^{8+} on Ne is not reached here. Also a strongly increased damping appears with decreasing velocity, nearly reaching the nonresonant case at 4.7 MeV collision energy. At this energy of 4.7 MeV, however, the angular resolution may be not quite good enough to clearly identify maxima and minima in $P_K^{Ar}(b)$. But even here the oscillations are still seen at small b and are completely damped out at large b at this very low impact energy.

With relation $P_K^{Ar} = 4P_1(1-P_1)\sin^2\Phi$, where P_1 is the transfer probability for single passage through the coupling region at R_c, and Φ is the phase accumulated between the two coupling regions around R_c [52]: $2\Phi = \hbar^{-1}\int(E_{1s\sigma} - E_{2p\sigma})\,dt$, with $E_{1s\sigma}(R(t))$ and $E_{2p\sigma}(R(t))$ binding energies of the electron in the 1sσ and 2pσ molecular states, respectively. Here, one assumed the collision is slow enough so that molecule-like orbitals with varying binding energies as function of the internuclear separation $R(t)$ are formed from the atomic orbitals. There are several simplifications in this approximation for Φ which are discussed in [52]. One of the simplifications is the localized coupling region at R_c. In Figure 8.27, this phase integral is illustrated with a S-Ar correlation diagram. It is scaled from H^+ and H to S^{15+} and Ar° (dashed curve) by adjusting to asymptotic Hartree–Fock energies of S^{7+} and Ar^{8+} in order to take care of the charge equilibration during the collision (full curve) [52]. The two localized coupling regions at R_c are indicated in Figure 8.27. The values of the phase angle Φ can be easily identified at the extrema in Figure 8.26. As $\Phi = (n - 1/2)\pi$, then the maxima are at $n = 1, 2, \ldots$ and the minima at $n = 3/2, 5/2, \ldots$.

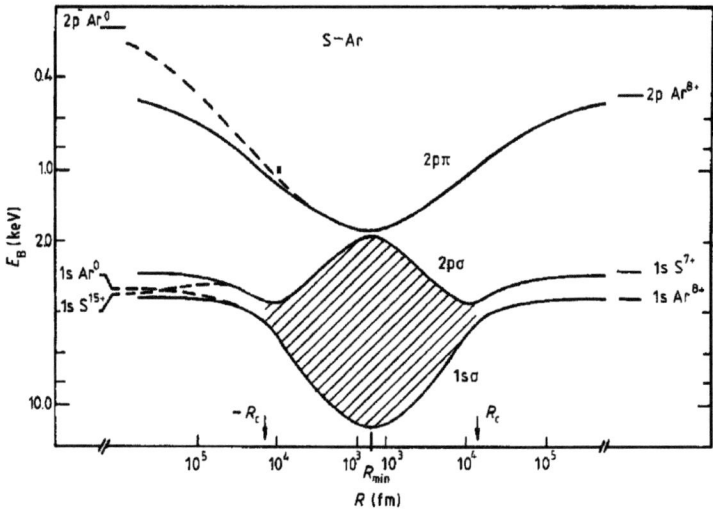

Figure 8.27: S-Ar correlation diagram [52] (for explanation of full and broken curves, see text).

It is shown in [52] that the b position of the maxima and minima, scaled with the ion velocity, indeed form a common curve. The analysis demonstrates [52] that one gets information about the $1s\sigma$-$2p\sigma$ molecular energy difference even from a quite complex and dynamic collision process. This approach opens perspectives for a spectroscopy of super heavy quasimolecules where the $1s\sigma$ binding energies can exceed the rest masses of electron-positron pairs and the diving of the bound states in the negative Dirac sea is still an open question.

Bibliography

[1]   Bohr N, On the Theory of the Decrease of Velocity of Moving Electrified Particles on passing through Matter. Phil. Mag. 1913;25:10–31.

[2]   Bohr N, Lindhard J. Dan. Mat. Fys. Medd. 1954;28(7).
[3]   Oppenheimer JR. Phys. Rev. 1928;31:349.
[4]   Brinkman HC, Kramers HA. Proc. K. Ned. Acad. Sci. 1930;33:973.
[5]   Ullrich J, Moshammer R, Dorn A, Dörner R, Schmidt LPhH, Schmidt-Böcking H. Rep. Prog. Phys. 2003;66:1463.
[6]   Dörner R, Mergel V, Jagutzki O, Spielberger L, Ullrich J, Moshammer R, Schmidt-Böcking H. Phys. Rep. 2000;330:95–192.
[7]   McDowell MRC, Coleman JP. Introduction to the Theory of ion Atom-Collisions. Amsterdam: North-Holland; 1970. Chapters 6–8.
[8]   Shakeshaft R, Spruch L. Rev. Mod. Phys. 1979;51:369.
[9]   Briggs JS, Macek JH. Adv. in Atomic, Molecular, Optical Phys. 1990;28:1.
[10] Bohr N. K. Dan Vidensk. Selsk. Mat.- Fys. Medd. 1948;XVIII:8.
[11] Chesire M. Proc. Phys. Soc. (London). 1964;84:89.
[12] Belkic Dz, Gayet R, Salin A. Phys. Rep. 1979;56:279.
[13] Chan FT, et al. Phys. Rev. A. 1979;20:1841.
[14] Dewangan DP, Eichler J. J. Phys. B. 19 (1986) 2939.
[15] Macdonald JR, Cocke CL, Edison WW. Phys. Rev. Lett. 1974;32:64.
[16] Nikolaev VS. Zh. Eksp. Teor. Fiz. 51, 1263 (1966), Sov. Phys. JETP. 1967;24:847.
[17] Thomas LH. Proc. Roy. Soc. (London), Ser. A 1927;114:561.
[18] Drisko BM. Ph. D. thesis, Carnegie Institute of Technology, 1955. Shakeshaft R. J. Phys. B. 1974;7:1059.
[19] Dettmann K, Leibfried G, Z. Phys. 1969;218:1.
[20] Kramer PJ. Phys. Rev. A. 1972;6:2125.
[21] McGuire H, Eichler J, Simony PR. Phys. Rev. A. 1983;28:2104.
[22] Macek, Alston S. Phys. Rev. A. 1982;26:250.
[23] McGuire JH, Sil NC. Phys. Rev. A. 1983;28:3679.
[24] McGuire JH, Kletke RE, Sil NC. Phys. Rev. A. 1985;32:815.
[25] Briggs JS, Greenland PT, Kocbach L. J. Phys. 1982;815:3085.
[26] Adivi EG, Bolorizadeh MA. J. Phys. B. 2004;37:3321.
[27] Abufager PN, Fainstein PD, Martínez AE, Rivarola RD, J. Phys. B. 2005;38:11.
[28] Humphries WJ, Moiseiwitsch BL, J. Phys. B 1984;17:2655.
[29] Schwab W, Baptista GB, Justiniano E, Schuch R, Vogt H. J. Phys. B. 1987;20:2825.
[30] Hvelplund P, Andersen A. Phys. Scr. 1982;26:375.
[31] Schryber U. Helv. Phys. Acta. 40 1023 (1967).
[32] Welsh LM, Berkner KH, Kaplan SN, Pyle RV. Phys. Rev. 1967;158:85.
[33] Williams JF. Phys. Rev. 1967;157:97.
[34] Toburen LH, Nakai MY, Langley RA. Phys. Rev. 1968;171:114.
[35] Berkner KH, Kaplan SN, Paulikas GA, Pyle RV, Phys. Rev. A 1965;729:140.
[36] Horsdal-Pedersen E, Cocke CL, Stockli M. Phys. Rev. Lett. 50, 1910 (1983).
[37] Vogt H, Schuch R, Justiniano E, Schulz M, Schwab W. Phys. Rev. Lett. 1986;57:2256.
[38] Toshima N. Phys. Rev. A. 1992;45:R2663.
[39] Schmidt HT, Cederquist H, Schuch R, Bagge L, Källberg A, Hilke J, Rensfelt K-G, Mergel V, Achler M, Dörner R, Spielberger L, Jagutzki O, Schmidt-Böcking H, Ullrich J, Reich H, Unverzagt M, Schmitt W, Moshammer R. Hyperfine Interact. 1997;108:339.
[40] Fischer D, Støchkel K, Cederquist H, Zettergren H, Reinhed P, Schuch R, Källberg A, Simonsson A, Schmidt HT. Phys. Rev. A. 2006;73:052713.
[41] Fischer D, Gudmundsson M, Berenyi Z, Haag N, Johansson HAB, Misra D, Reinhed P, Kallberg A, Simonsson A, Støchkel K, Cederquist H, Schmidt HT. Phys. Rev. A. 2010;81:012714.
[42] Voitkiv AB, Najjari B, Ullrich J. Phys. Rev. Lett. 2008;101:223201.

[43] Pálinkás J, Schuch R, Cederquist H, Gustafsson O. Phys. Rev. Lett. 1989;63:2464.

[44] Mergel V, Dörner R, Achler M, Khayyat Kh, Lencinas S, Euler J, Jagutzki O, Nüttgens S, Unverzagt M, Spielberger L, Wu W, Ali R, Ullrich J, Cederquist H, Salin A, Wood CJ, Olson RE, Belkić Dž, Cocke CL, Schmidt-Böcking H. Phys. Rev. Lett. 1997;79:387.

[45] Schmidt HT, Fardi A, Schuch R, Schwartz SH, Zettergren H, Cederquist H, Bagge L, Källberg A, Jensen J, Rensfelt K-G, Mergel V, Schmidt L, Schmidt-Böcking H, Cocke CL. Phys. Rev. Lett. 2002;89:163201.

[46] Gudmundsson M, Fischer D, Haag N, Johansson HAB, Misra D, Reinhed P, Schmidt-Boecking H, Schuch R, Schoeffler M, Støchkel K, Schmidt HT, Cederquist H. J. Phys. B. 2010;43:185209.

[47] Zeilinger A, Gähler R, Shull CG, Treimer W, Mampe W, Rev. Mod. Phys. 1988;60:1067.

[48] Carnal O, Mlynek J. Phys. Rev. Lett. 1991;66:2689.

[49] Gerlich S, Eibenberger S, Tomandl M, Nimmrichter S, Hornberger K, Fagan PJ, Tüxen J, Mayor M, Arndt M. Nature Com. 2011;2:263.

[50] Ziemba FP, Everhart E. Phys. Rev. Lett. 1959;2:299. Ziemba FP, Lockwood GJ, Morgan GH, Everhart E. Phys. Rev. 1960;118:1552.

[51] Barany A, Danared H, Cederquist H, Hvelplund P, Knudsen H, Pedersen JOK, Cocke CL, Tunnell LN, Waggoner W, Giese JP. J. Phys. B: A 1986;19:L427.

[52] Schuch R, Ingwersen H, Justiniano E, Schmidt-Böcking H, Schulz M, Ziegler F, J. Phys. B. 1984;17:2319.

[53] Stolterfoht N, Sulik B, Hoffmann V, Skogvall B, Chesnel JY, Rangama J, Frémont F, Hennecart D, Cassimi A, Husson X, Landers AL, Tanis JA, Galassi ME, Rivarola RD. Phys. Rev. Lett. 2001;87:023201.

[54] Misra D, Kelkar A, Kadhane U, Kumar A, Tribedi LC, Fainstein PD. Phys. Rev. A. 2006;74:060701(R).

[55] Wang YD, McGuire JH, Rivarola RD. Phys. Rev. A. 1989;40:3673.

[56] Tuan TF, Gerjuoy E, Phys. Rev. 1960;117(3):756.

[57] Deb NC, Jain A, McGuire JH. Phys. Rev. A. 1988;38:3769.

[58] Shingal R, Lin CD. Phys. Rev. A 1989;40:1302.

[59] Wang YD, McGuire JH, Rivarola RD. Phys. Rev. A. 1989;40:3673.

[60] Stochkel K, Eidem O, Cederquist H, Zettergren H, Reinhed P, Schuch R, Cocke CL, Levin SB, Ostrovsky VN, Kallberg A. Phys. Rev. A. 2005;72:050703(R).

[61] Cheng S, Cocke CL, Frohne V, Kamber EY, McGuire JH, Wang Y. Phys. Rev. A. 1993;47:3923.

[62] Corchs SE, Busnengo HF, Rivarola RD, McGuire JH, Nucl. Instr. Meth. B 1996;117:41.

[63] Schmidt HT, Fischer D, Berenyi Z, Cocke CL, Gudmundsson M, Haag N, Johansson HAB, Kallberg A, Levin SB, Reinhed P, Sassenberg U, Schuch R, Simonsson A, Støchkel K, Cederquist H. Phys. Rev. Lett. 2008;101:083201.

[64] Martínez S, Bernardi G, Focke P, Fregenal D, Suárez S. Phys. Rev. A. 2005;72:062722.

[65] Misra D, Schmidt HT, Gudmundsson M, Fischer D, Haag N, Johansson HAB, Kallberg A,Najjari B, Reinhed P, Schuch R, Schoffler M, Simonsson A, Voitkiv AB, Cederquist H. Phys. Rev. Lett. 2009;102:153201.

[66] Lamichhane BR, Arthanayaka T, Remolina J, Hasan A, Ciappina MF, Navarrete F, Barrachina RO, Lomsadze RA, Schulz M. Phys. Rev. Lett. 2017;119:083402.

[67] Schmidt LPhH, Schössler S, Afaneh F, Schöffler M, Stiebing KE, Schmidt-Böcking H, Dörner R. Phys. Rev. Lett. 2008;101:173202.

[68] Schmidt LPhH, Lower J, Jahnke T, Schößler S, Schöffler MS, Menssen A, Leveque C, Sisourat N, Taïeb R, Schmidt-Böcking H, Dörner Phys R. Rev. Lett. 2013;111:103201.

[69] Bohr N. Albert Einstein: Philosopher Scientist. Cambridge, England: Cambridge University Press; 1949. p. 201.

[70] Schuch R. In: Datz S, editor. XII ICPEAC book of invited papers. Amsterdam: North Holland

publ.; 1982. p. 151.

[71] Lichten W. Phys. Rev. 1963;131:229; Phys. Rev. A. 1965;27:139.

[72] Taulbjerg K, Briggs J, Vaaben J. J Phys B 1976;9:1351.

[73] Nolte G, Volpp J, Schuch R, Lichtenberg W, Schmidt-Böcking H, Specht HJ. Journ. Phys. B. 1980;13:4599.

[74] Ziemba P, Russek A. Phys. Rev. 1959;115:922.

[75] Sepp W-D, Fricke B, Morović T. Phys. Lett. A. 1981;81:258.

[76] Soff G, Muller B, Greiner W. Phys. Rev. Lett. 1978;40:540.

[77] Liesen D, Armbruster P, Bosch F, Hagmann S, Mokler P, Schmidt-Bocking H, Schuch R, Wilhelmy J. Phys. Rev. Lett. 1980;44:983.

[78] Schuch R, Nolte G, Schmidt-Böcking H, Lichtenberg W. Phys. Rev. Lett. 1979;43:1104.

[79] Hagmann S, Cocke CL, Macdonald JR, Richard P, Schmidt-Böcking H, Schuch R. Phys. Rev. A. 25, 1918 (1982).

[80] Lin CD, Tunnel L. Phys. Rev. A. 22, 76 (1980); Lin CD, Winter TG, Fritsch W. Phys. Rev. A. 1982;25:2395.

A. Cassimi, X. Fléchard, B. Gervais, A. Méry, and J. Rangama

9 Ion-dimer collisions

1 Introduction

Dimer targets constitute the simplest intermediate state from gas phase to condensed matter. Their study is of growing interest as understanding multi-charged ion (MCI) collisions involving atomic dimers may help to pinpoint differences between the processes at play with an isolated target and those occurring when this target is embedded in an environment. In other words, dimers bear this quite unique potential to provide a bridge between a few-body problem, with one projectile and a simple atomic target, and a many-body problem where only collective quantities can be obtained. Fundamental aspects of electronic excitation and relaxation toward atomic motion leading to defects in structural reorganization and chemical reactions become accessible. The investigation of MCI dimer collisions can thus lead to discoveries that are crucial for the modeling of a large variety of systems in the fields of astrophysics, atmospheric science, radiation damage, or even hadron-therapy.

Rare-gas dimers are linked by van der Waals bonds originating from correlations in the fluctuating polarizations of neighbor particles. Compared to ionic or covalent bonds, van der Waals bonds are much weaker and associated to larger internuclear or intermolecular distances. As an extreme example, one can take the He_2 dimer, with an average bond length and an average binding energy of respectively 98 a. u. and 95 neV [1]. What should we expect for such weakly bound systems, made of two nearly independent atoms, when excited by light or by the impact of charged particles such as MCIs? The goal of this chapter is to present a brief overview of the results gathered along the last decade in collisions between MCI projectiles and dimer targets.

If atomic and molecular dimer targets are obviously more complex than their constituting monomers, these targets remain simple enough to enable kinematically complete measurements using the so-called recoil-ion momentum spectroscopy technique [2, 3]. We will concentrate on experiments making use of this technique, for which most of the data were obtained in the low velocity regime (with a perturbation of Sommerfeld parameter $\eta \gg 1$, see Chapter 2). In this regime, electron transfer from the target to the projectile ion is by far the dominant primary process leading to the ionization of the target (see Chapter 7). Other experiments, focusing on radiation damage, were also performed with projectile energies corresponding to the

A. Cassimi, B. Gervais, A. Méry, J. Rangama, CIMAP, CEA-CNRS-ENSICAEN-UNICAEN, Normandie Université, BP5133, F-14050 Caen Cedex 04, France, e-mails: cassimi@ganil.fr, gervais@ganil.fr, mery@ganil.fr, rangama@ganil.fr
X. Fléchard, ENSICAEN, UNICAEN, CNRS/IN2P3, LPC Caen, Normandie Univ, 14000 Caen, France, e-mail: flechard@lpccaen.in2p3.fr

https://doi.org/10.1515/9783110580297-009

maximum of the stopping power in liquid water. As will be shown, these recent measurements provide key information on how the collision energy is deposited in the system and how the latter relaxes and eventually affects the neighboring environment.

In a first section, we will briefly describe the basic principles of the experimental technique used to study collisions between MCIs and dimer targets. A second section will focus on the main results obtained during the last decade with rare-gas dimer targets. We will discuss the differences observed in the collision dynamics and the subsequent fragmentation for van der Waals bound targets when compared to covalent bound targets. In particular, the appearance of relaxation processes resulting from the proximity of a second atomic center will be highlighted. In the third section, calculations performed using a classical model, the Monte Carlo classical over-the-barrier model (MC-COBM) [16], developed for atomic dimer targets, will be presented and confronted to experimental data. A short and last section will discuss possible perspectives in the field of ion-dimer collisions.

2 Experimental technique

The experimental results discussed in this chapter were all obtained using the so-called recoil-ion momentum spectroscopy (RIMS) technique. Its principle is based on the use of a weak electric field[1] to extract the recoiling ions or fragments from the target towards a position sensitive detector (PSD) as soon as those low energy charged particles are produced. The instant of the collision is inferred here either from the detection time of the scattered projectile, whose charge can be analyzed using deflection plates, or by a time signal from the beam buncher when using pulsed projectile beams. The charge and full momentum vector of the recoil particles are then inferred from their time of flight (TOF) and position on the detector. When dealing with molecular targets, the recoil ion PSD and its data acquisition system must provide multihit capability with a short dead time and enable the detection of several fragments from the same target. As the target velocity spread is crucial for this technique, it requires a cold target that is usually provided by a supersonic gas jet. The RIMS technique is often complemented by another PSD for electrons, combined with a weak magnetic field to confine their trajectories within the range of the detector. The setup is in that case called a reaction microscope and gives access to the complete kinematics of the collision when electrons are also emitted. As comprehensive reviews on these techniques can be found in Chapters 2 and 3 or previous articles [2, 3], we will only give a brief description of their application to dimer targets.

---

1 For experiments using dimer targets, the electric field is typically a few V/cm.

Dimer targets are provided within a cold supersonic gas jet of ~1 mm diameter crossing the projectile beam at 90° at the center of the recoil-ion spectrometer. They are produced during the supersonic expansion of the gas jet by optimizing parameters such as the stagnation gas pressure (of the order of a few bars), the initial gas temperature,[2] and the nozzle diameter (typically a few tens of µm) [4, 5, 6]. The dimer production represents only a small fraction (< 1%) of the monomers in the jet target, and a large contribution of random coincidences due to collisions with monomers needs to be suppressed. An efficient filter can be applied if the dimer fragments into only charged fragments after the collision. In a first step, the charge and mass of each fragment are identified thanks to their respective TOF. As the kinetic energy released (KER) due to Coulombic explosion is very large compared to the energy transfer induced by the collision, the latter can in a first step be neglected.[3] By simply imposing $\mathbf{P_1} + \mathbf{P_2} = \mathbf{0}$, the momentum vectors of the first and second fragments $\mathbf{P_1}$ and $\mathbf{P_2}$ can be calculated in the laboratory frame using their TOFs and positions on the detectors with the initial coordinates of the fragmentation as free parameters. By doing so, $\mathbf{P_1}$ and $\mathbf{P_2}$ give the exact momenta of the two fragments in the frame of the center of mass and an approximate initial 3D position of the fragmentation can be reconstructed. As shown in [7] and Fig. 9.1 for the fragmentation of $(Ar_2)^{3+}$ into $Ar^{2+} + Ar^+$ fragments, a selection of reconstructed positions landing close or within the volume of the collision region suppress drastically the background mainly due to monomers. For each selected event, $\mathbf{P_1}$ and $\mathbf{P_2}$ can then simply be used to determine the fragmentation kinetic energy released,

$$\text{KER} = \frac{P_1^2}{2m_1} + \frac{P_2^2}{2m_2} \tag{9.1}$$

and the orientation of the molecule with a resolution that is not limited, to first order, by the size of the collision region. Momenta $\mathbf{P_1}$ and $\mathbf{P_2}$ can give access to the orientation of the molecule at the instant of the collision when the molecule rotation is much slower than the fragmentation process.[4] Knowing the orientation of the molecule is then particularly valuable for the study of processes depending on the angle between the molecule and the projectile beam axis. On the other hand, within the so-called reflection approximation, the KER gives direct access to the internuclear distance R at the instant when finally both sites are ionized. The Coulomb repulsion between the two sites is the dominant contribution and, in the case of rare-gas dimers, it can be approximated as KER $= q_1 q_2/R$ (in atomic units), with q_1 and q_2 the charges of the two fragments.[5] Combined with the calculation of theoretical potential energy curves

---

2 Some rare gas species such as He require precooling.

3 The KER due to Coulombic explosion is typically of the order of one eV and more while the energy transferred by the projectile to the recoil molecular ion represents only a few meV.

4 It is the case if no metastable transient states are populated prior fragmentation.

5 For a precision level of a few percent, the polarizability and other correction terms can be neglected.

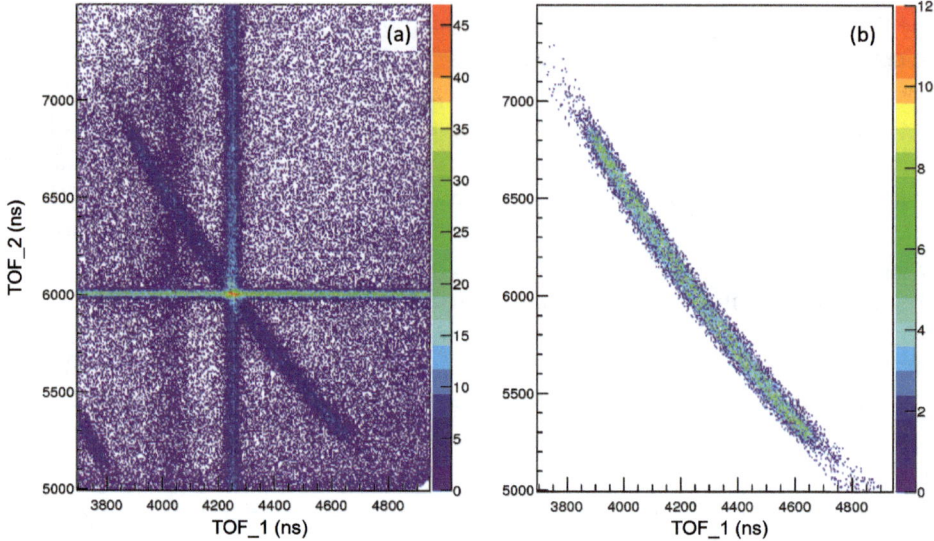

Figure 9.1: TOF coincidence map (second fragment TOF versus first fragment TOF) around the Ar^{2+}–Ar^+ island with all the double-hit events (a) and after applying a 3D momentum conservation condition (b). Counts are given by the color scale. The intense vertical and horizontal lines in (a) are due to false coincidences where two projectiles capture electrons from distinct Ar atoms in the supersonic jet.

(PECs) of the dimer ion, KER spectra are then very helpful for the identification of the transient states populated by the collision and of the relaxation processes at play.

One can additionally determine the vector momenta P'_1 and P'_2 in the laboratory frame, assuming now that the fragmentation takes place at the exact center of the collision region. The recoil energy of the target is then no more neglected and the total momentum $P'_1 + P'_2$ gives access, although with limited resolution, to the momentum exchange between the projectile and the target.[6] As will be shown in the following sections, this momentum exchange can be directly linked to the projectile scattering angle, providing further insights into the collision process and additional points of comparison with theory.

Despite the difficulty to produce a clean dimer target, the RIMS technique is thus particularly powerful when the target ion relaxes its excitation energy by fragmentation into only charged fragments. The charge and mass of each fragment provide an identification of the fragmentation channels, momentum conservation allows for a clean selection of the events, and the momenta of the fragments give access to the KER, to the initial orientation of the dimer and to the momentum exchange between

---

6 $P'_1 + P'_2$ can differ from zero both because of the momentum transferred during the collision and because of the spatial extension of the collision region. The initial momentum of the target is negligible for most experiments of this chapter.

the projectile and the target. Moreover, these quantities can be measured in coincidence with the charge of the scattered projectile in experiments where the latter are detected, and with the energy of electrons potentially emitted by the system when using a complete reaction microscope.

3 Experimental results in collisions with rare-gas dimers

Rare-gas dimers have attracted much attention as they allow the study of a situation where the system can be considered as two almost independent but nevertheless weakly bound neighbor atoms. In particular, photon ionization experiments [8, 9] have confirmed the predictions of Cederbaum et al. [10] showing that new relaxation processes such as interatomic Coulombic decay (ICD) can open-up in the vicinity of a neighbor atom. ICD takes place when the excitation energy of one of the atom is not sufficient to relax through Auger decay but large enough to trigger the ionization of the neighbor atom. Despite the large distance between the two sites, the latter process, through the exchange of a virtual photon, can occur (Figure 9.2).

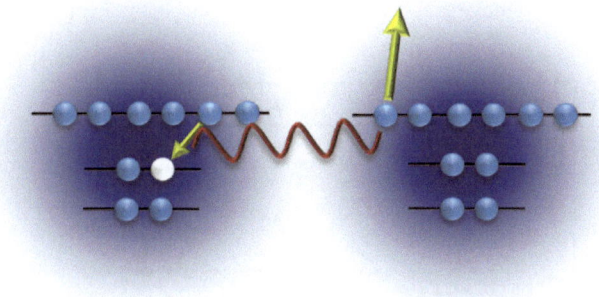

Figure 9.2: Illustration of the ICD process for a neon dimer: the decay toward an inner shell vacancy leads to the ionization of the neighbor site via the exchange of a virtual photon.

But photon ionization can barely produce direct multiple ionization, and the production of higher charge molecular ions relies on subsequent relaxation processes driven by the electron-electron correlation, such as Auger emission or ICD. By contrast, in collisions involving relatively slow MCI projectiles, the large intensity of the Coulomb field can lead to multiple processes with a high probability and produce a multiply ionized target with a limited degree of excitation [11]. A dimer target can then be ionized by removing electrons either from one of the atoms or from both of them and one needs to elucidate both the dynamics of electron capture and of the subsequent dissociation. When one site of the dimer is highly ionized, what kind of charge rearrangement can be expected with the neighbor atom? Would a slow or limited electron mobility between the two atomic sites allow to investigate how the electrons are captured by the MCI projectile? And for low degrees of ionization, does the molecular ion relax through the same new processes as observed when using photon ionization? First, answers were obtained in recent experiments performed at the ARIBE and SME beamlines of GANIL[7] in Caen and at the Van de Graaff accelerator of the Institut für Kernphysik of the Gothe-Universität in Frankfurt.

3.1 Low electron mobility in dimer targets

The first experimental results were obtained for low-energy (152 keV) collisions between Ar^{9+} projectiles and Ar_2 dimer targets [4]. The Ar_2 bond length of 7.2 a. u. is here significantly larger than in covalent molecules (2.35 a. u. for N_2). In this experiment, up to 4 electrons were removed from the target by charge transfer, leading to the fragmentation channels $(q_1; q_2)_F$, where q_1 and q_2 are the charges of the Ar fragments (the subscript F standing for fragmentation), and to the respective branching ratios br given in Table 9.1.

Table 9.1: Fragmentation channels and their respective experimental branching ratio obtained in $Ar^{9+} + Ar_2$ collisions.

Relaxation reaction	Fragmentation channel	Branching ratio br
$(Ar_2)^{2+} \rightarrow Ar^+ + Ar^+$	$(1;1)_F$	0.42
$(Ar_2)^{3+} \rightarrow Ar^{2+} + Ar^+$	$(2;1)_F$	0.30
$(Ar_2)^{4+} \rightarrow Ar^{3+} + Ar^+$	$(3;1)_F$	0.19
$(Ar_2)^{4+} \rightarrow Ar^{2+} + Ar^{2+}$	$(2;2)_F$	0.09

The most remarkable result when looking at branching ratios is the preference for the asymmetric channel $(3;1)_F$ over the more energetically favorable symmetric channel $(2;2)_F$ following quadruple electron capture. This was found in sharp contrast with the fragmentation of diatomic covalent molecules, for which symmetric charge sharing is by far dominant [12, 13, 14, 15]. Such a reverse order of priority in charge sharing for homonuclear diatomic system is indeed characteristic of van der Waals dimers: for covalent molecules, a fast rearrangement of the charges following the collision favors

---

7 ARIBE and SME are experimental areas for very low energy ion beams and intermediate energy ion beams at the Grand Accélérateur National d'Ions Lourds (GANIL).

the symmetric fragmentation channels, but the fragmentation channels of rare-gas dimers still reflect the primary collision process that may favor electron capture from one of the two sites. These results were thus a strong indication for, on one hand, the presence of asymmetric capture processes, and on the other one, of a low electron mobility[8] within the resulting dimer ion.

The low electron mobility between the two sites of the dimer is correlated to the shapes and positions of potential energy curves (PEC) of the dimer ions. The PECs shown in Figure 9.3 represent the potential energy of the dimer ion as a function of the internuclear distance R for several molecular states that may be populated by multiple electron capture. In the following, the primary capture processes are noted $(q_1; q_2)_C$, q_1, and q_2 being the numbers of electrons removed from each site (the subscript C standing for capture). For q_1 and $q_2 > 0$, the Coulomb repulsion instantly triggers the fragmentation in the $(q_1; q_2)_F$ channel. Only for the most asymmetric capture channels $(3; 0)_C$ (PEC $Ar^{3+} + Ar^0$) and $(4; 0)_C$ (PEC $Ar^{4+} + Ar^0$), curve crossings with excited states located close to the equilibrium internuclear distance (7.2 a. u.) can lead to charge transfer (CT) between the two sites, and thus, to the more symmetric fragmentation channels $(2; 1)_F$ and $(3; 1)_F$, respectively. Except for this little charge rearrangement, the fragmentation channel reflects thus the initial capture process. Finally, the asymmetric channel $(Ar^{2+} + Ar^0)$ populated by 1-site double capture does not show any crossing with another PEC. The only pathway open for de-excitation toward the dissociative ground state of the dication (PEC $Ar^+ + Ar^+$), and thus to the fragmentation channel $(1; 1)_F$, is through the RCT (radiative charge transfer) process. As shown in Figure 9.3, this relatively slow process[9] involving photon emission is expected to take place at shorter internuclear distance than the equilibrium distance (the so-called Franck–Condon region shown in gray). It results in a larger KER in the $(1; 1)_F$ dissociation channel than for a 2-site double capture triggering a direct Coulomb explosion (Figure 9.4).

Again, thanks to the absence of crossings, the information on the primary collision process is kept and the 1-site versus 2-site double capture contributions can be inferred from the experimental KER spectrum, as shown in Figure 9.4. The same features, i. e., the predominance of asymmetric channels for high capture multiplicity and the relaxation through RCT for 1-site double capture, have been observed for all the low-energy collision systems involving rare-gas dimers investigated so far [16, 18]. It is a strong indication that low electron mobility may be a generic property of ionized rare-gas dimer targets. One can see the main interest of this specific property: as the primary collision process is not washed out by electronic rearrangement, relative cross-sections associated to all possible capture multiplicities on both atomic sites be-

---

8 We speak about low electron mobility as a result of the strong localization and binding of the electrons around their respective nuclei.

9 The typical lifetime for the RCT process is of the order of the *ns*.

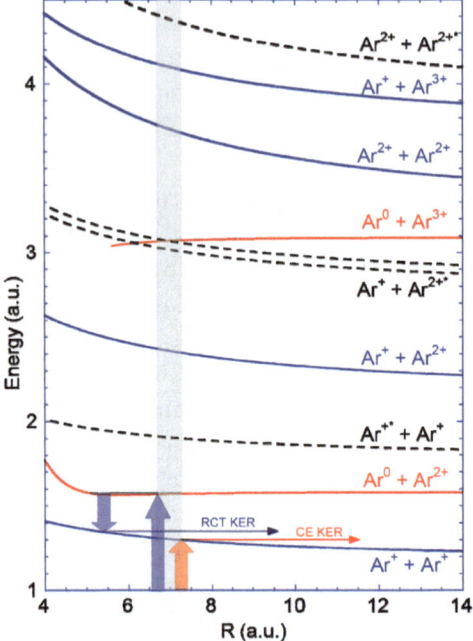

Figure 9.3: Approximate potential energy curves for different fragmentation channels of $(Ar_2)^{2+}$ and $(Ar_2)^{3+}$ obtained by asymptotic limits, Coulombic and polarization energies. Solid (dashed) lines correspond to channels with fragments in their ground (first excited) states. The light and dark gray arrows represent the Coulomb explosion (CE) and radiative charge transfer (RCT) processes, respectively. The gray band shows the position of the Franck–Condon region. (Reproduced with permission from [4].)

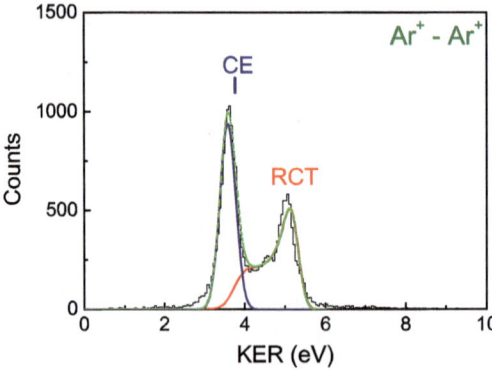

Figure 9.4: KER distributions of the fragment pair $Ar^+ + Ar^+$. The shape of the CE (blue line) and RCT (red line) contributions were obtained theoretically [4].

come accessible, providing new insights into the dynamics of a collision with a 2-site target.

3.2 Access to the primary collision processes

As discussed in the previous paragraph, at low energy, relative capture cross sections of the primary collision processes may be extracted thanks to the low electron mobility within the ionized dimer. If the fragmentation channels and KER spectra that are provided directly by the experiment mainly reflect the primary capture multiplicity on each site, some corrections remain to be made. The main correction is simply due to the fact that the experimental technique is blind to dissociation channels comprising one neutral atom. For a 1-site capture of q electrons $(q; 0)_C$, transient non-dissociative states will be populated. Those will eventually end up in the nearest dissociative channel $(q - 1; 1)_F$ through RCT or CT. However, dissociative states may also be populated, leading to a direct fragmentation $(q; 0)_F$, with one neutral atom. Such events are lost to the experiment and, as discussed in more details in [16], getting a good estimate of their probability is not trivial. In [17, 18], relative cross sections were estimated by considering that dissociative and nondissociative states were populated in equal amount. The resulting cross-sections were found in good agreement with the MC-COBM (Monte Carlo – classical over-the-barrier model) presented in the next section. In a later study, two scenarios were considered and compared: one with an equal population of dissociative versus nondissociative, another with only nondissociative states. Both scenarios led to quite similar results and have shown reasonable agreement with the model [16]. Providing that the corrections discussed above can be performed with an acceptable accuracy, low electron mobility within dimer ions allows indirect access to charge exchange relative cross-sections for comparison with theory. Nevertheless, the sensitivity to neutral fragments remains the main weakness of the experimental technique. Both theoretical and experimental developments will thus be needed if one wants to improve the precision on relative cross-sections for primary charge exchange process.

Another manifestation of the low electron mobility within dimers was found when investigating the correlation between the projectile scattering angle and the orientation of the fragmenting dimer ion. Information on the scattering angle of the projectile can be particularly helpful here, as at low energy, it is strongly related to the impact parameter associated to the collision process.

A recent study, conducted with Ar^{9+} low energy (135 keV) projectiles and Ar_2 dimer targets, has shown a strong correlation indicating that the projectile is preferentially scattered in the same direction as the fragment carrying the higher charge [17]. Such correlations were observed by measuring in coincidence the angles in the azimuthal plane of the most charged fragment, $\phi_{Ar^{A+}}$, and of the scattered projectile, ϕ_{proj}, as illustrated in Figure 9.5. This result is once again quite the opposite of what had been previously observed for very low energy collisions with N_2 covalent molecules [19]. A charged projectile generates a highly localized electric field, and electron capture or ionization from the near-site atom is expected to be favored. With covalent molecules, a fast intramolecular charge redistribution can occur and Coulomb repulsion between the ionized target and the projectile ion may lead to a lower charge on the near-site, as

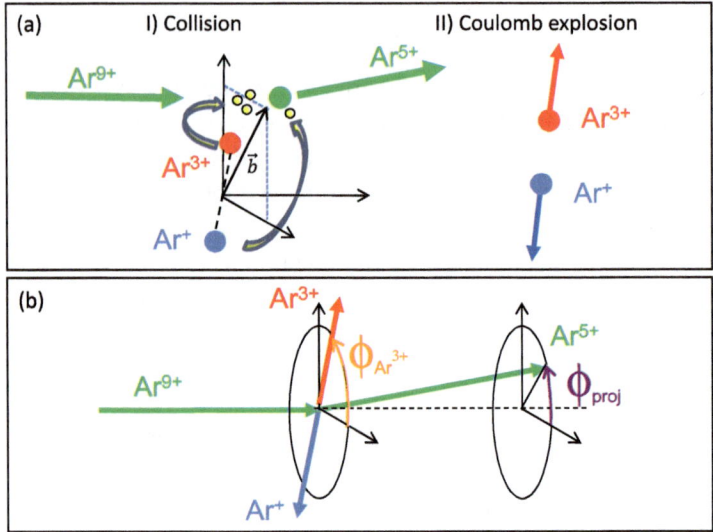

Figure 9.5: Schematic view of the multiple electron capture from Ar_2 by Ar^{9+} projectiles resulting in the $Ar^{3+} + Ar^+$ asymmetric fragmentation channel (a). Representation of the scattering angle ϕ_{proj} and of the angle of emission $\phi_{Ar^{3+}}$ of the most charged fragment in the plane transverse to the beam axis (b). (Reproduced with permission from [17].)

observed in [19]. By contrast, the low electron mobility within dimers should prevent this charge rearrangement so that the correlation between the projectile scattering angle and the direction of the most charged fragment is conserved. Figure 9.6 shows the angular distributions $(\phi_{proj} - \phi_{Ar^{A+}})$ between scattered projectiles and fragments, in the plane transverse to the projectile beam axis, obtained by Iskandar et al. [17]. For the asymmetric fragmentation channels $(Ar^{2+} + Ar^+)$ and $(Ar^{3+} + Ar^+)$ noted $(2;1)_F$ and $(3;1)_F$ respectively and resulting from triple capture (TC) and quadruple capture (QC), the most charged fragment Ar^{A+} serves as a reference, with an angle set at 0°. For double capture (DC), 1-site and 2-site capture channels (noted respectively $(2;0)_C$ and $(1;1)_C$) were analyzed independently. However, as 1-site DC ends up in the symmetric fragmentation $(Ar^+ + Ar^+)$, the memory of the site ionized during the primary collision process is lost, leading to a symmetric angular distribution. The experimental angular correlations were compared to calculations performed using a classical model MC-COBM whose description will be given in the next section. Because of the limited experimental resolution on the projectile scattering angle (see Section 2), the calculations were also convoluted with the instrumental function response. The very good agreement between calculations and experimental data shown in Figure 9.6 demonstrates that this model gives a realistic picture of the multiple capture process with structured targets such as dimers.

As the model provides direct access to the impact parameter $p(\boldsymbol{b})$ in the molecular frame, the 2D maps of $p(\boldsymbol{b})$ associated to each process were produced, confirming that

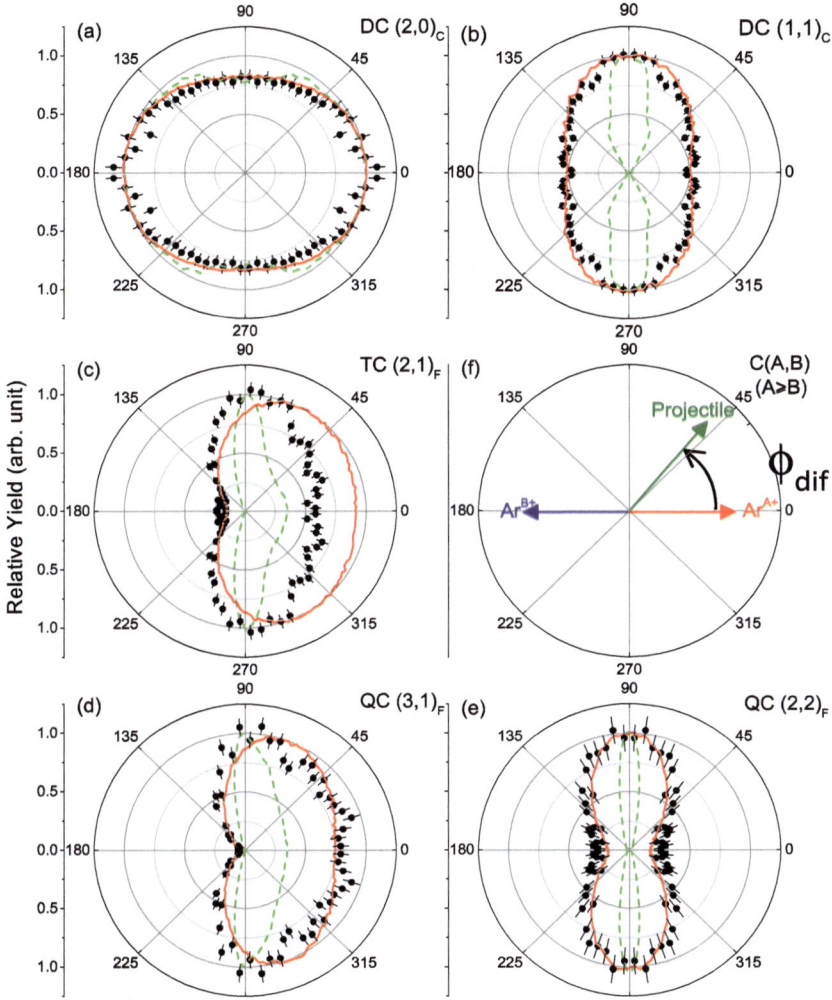

Figure 9.6: Projectile-fragment angular correlations for the DC (a) and (b), TC (c), and QC (d) and (e) and their associated fragmentation channels. Experimental data (black dots) are compared to the calculations with (red lines) and without (dashed green lines) convolution with the experimental function response. (Reproduced with permission from [17].)

capture from the near-site is strongly favored [17]. An example is given in Figure 9.7 for 1-site DC. For the first time, this joint experiment/model study could thus provide access to atomic site sensitivity in low energy collisions between MCIs and a complex target.

Collisions at intermediate energies (150 keV/u) were also investigated using He^{2+} projectiles and He dimer targets [6]. In this collision regime where the velocity of the projectile is comparable to the classical velocity of the target active electrons, charge

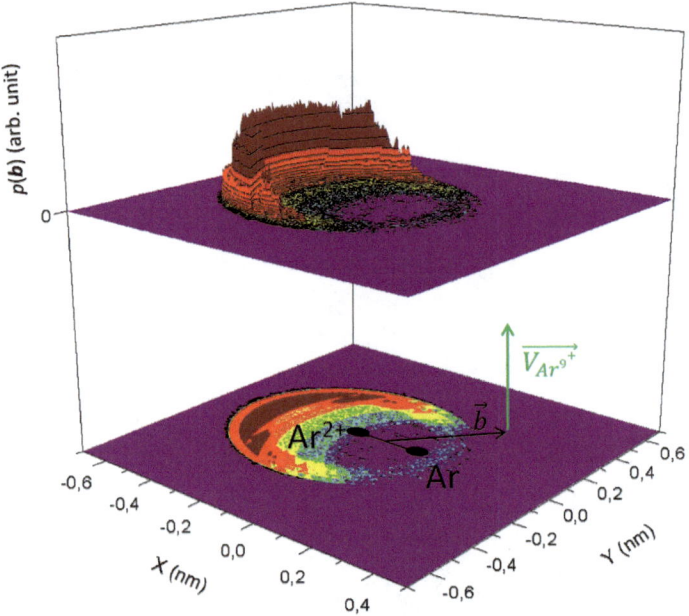

Figure 9.7: Probability map $p(\boldsymbol{b})$ obtained for the 1-site capture channel (2;0)C. The dimer is aligned on the x-axis and its center of mass located at coordinates (0;0). The projectile velocity is chosen transverse to the (x, y) plane. The color scale gives the probability $p(\boldsymbol{b})$ for this capture channel resulting in the removal of two electrons from one site. It clearly shows that impact parameters close to the Ar^{2+} site are strongly favored.

exchange is not anymore the dominant process and excitation as well as ionization of the target start to compete. By measuring the charge of the scattered projectile for the target dissociation in two He^+ fragments, two final channels could be distinguished: DC, with two electrons transferred from the target to the projectile, and TI (transfer ionization), with one electron transfer and one electron emitted into the continuum. The experiment was performed using a reaction microscope enabling the detection of electrons and the measurement of their energy. The fragment KER distributions associated to the DC and TI channels are displayed in Figure 9.8(a).

For the TI channel, three different peaks in KER could be observed and attributed to three different processes illustrated in Figure 9.8(b) to 9.8(d). The first structure at low KER (from 1 eV to 5 eV) corresponds to a 2-step process, with one single capture (SC) from one site and one single ionization (SI) from the other site. Such a process triggers direct dissociation of the dimer ion occurring at internuclear distances ranging approximately from $R = 5$ a. u. to 30 a. u. (within the reflection principle approximation). This is quite large compared to the impact parameters leading to SC and SI and we can now talk of a sequential 2-step process, contrasting with the 2-site process discussed previously for smaller dimers such as Ar_2. Peaks 2 and 3, at larger KERs, can be attributed to dissociations of the dimer ion occurring at smaller internuclear

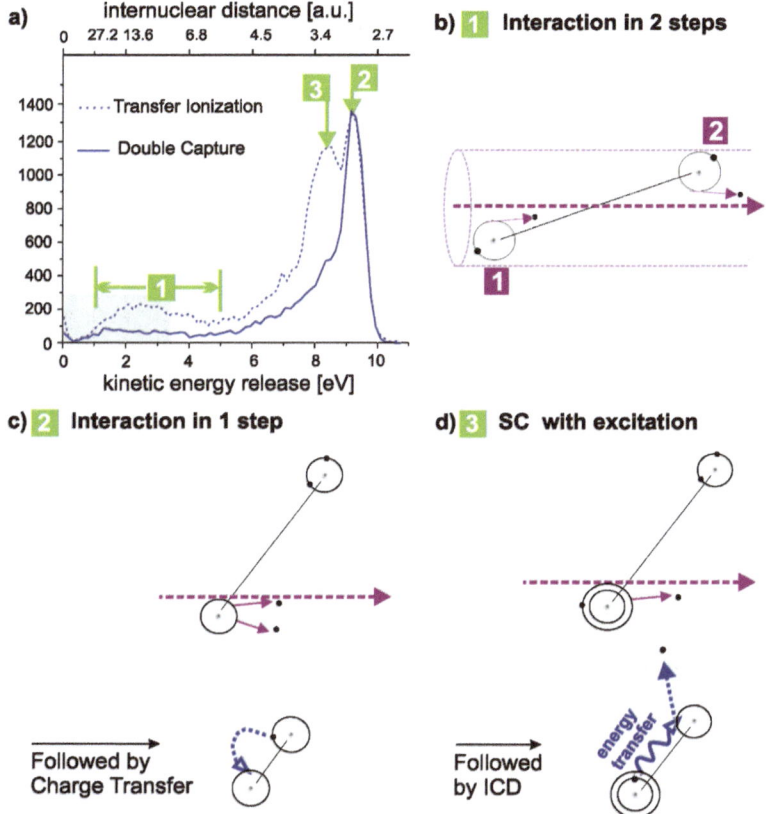

Figure 9.8: (a) Distribution of the KER after breakup into $He^+ + He^+$ for DC and TI. (b)–(d) Sketches of the sequence of events leading to the different peaks in (a); see text. (Reproduced with permission from [6].)

distances. An identification of the corresponding processes could be obtained thanks to PECs calculations and to the kinetic energy spectrum of the electrons. Figure 9.9 shows the kinetic energy of continuum electrons for TI as a function of the KER of the fragments. For events resulting in the KER structure peaked at about 8.5 eV, the kinetic energy of the emitted electron is correlated to the KER of the fragments, with diagonal patterns clearly visible in Figure 9.9. This is the typical signature of an ICD process [8] following electron capture from one center of the dimer along with an excitation of the remaining electron towards shells $n = 2, 3$ or 4. After the collision process, as the ion dimer shrinks toward shorter internuclear distances, the ICD process can occur close to the crossings between the transient $(He^+(n) + He)$ and the dissociative $(He^+ + He^+)$ PECs. Energy conservation then leads to the specific correlation between the KER and electron kinetic energy, according to the equation also displayed in Figure 9.9. This was the first experimental result showing direct evidence for the ICD process in collisions

$$E_{a,ICD} = 54{,}42 - 13{,}6 \times 4/n^2 - (2 \times 24{,}59 - KER)$$

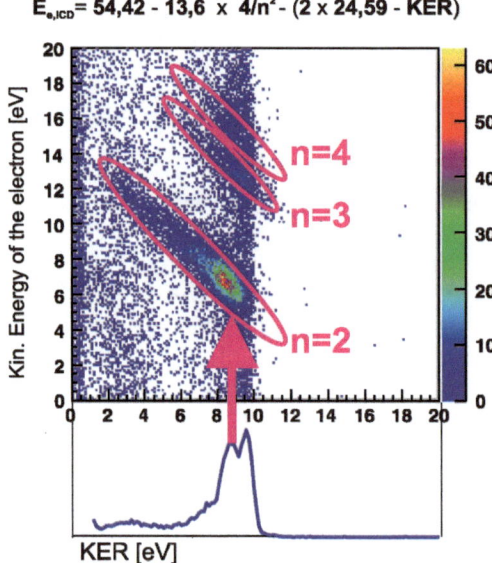

Figure 9.9: Kinetic energy release (ions) versus kinetic energy of the continuum electron for TI. The diagonal feature is a characteristic of ICD (equation in the top). $n = 2, 3, 4$ indicates the ICD electrons for excitation to the intermediate state [He + (n), He]. (Reproduced with permission from [6].)

with MCIs. Peak 2 is attributed to the RCT process: single capture and single ionization from the same site of the dimer populates the transient bound states of (He^{2+}–He), whose calculated potential energy curves show a minimum of the potential energy located at about 3 a. u. [6]. These states thus naturally decay toward the fragmentation channel $(1+;1+)_F$ through the RCT process at internuclear distances close to 3 a. u., yielding the KERs of peak 2 at about 9 eV. They are not associated with any structure in the electron spectrum as ionization is only caused by the collision primary process.

Similarly, for the DC channel, the peak 1 of Figure 9.8(a) can then simply be attributed to a 2-step process with one SC on each site. Peak 2 also results from the population of (He^{2+} – He) states through a TI process with one single site of the dimer. As the ICD process triggers the emission of an electron into the continuum, its contribution disappears from the KER spectrum of the DC channel in Figure 9.8(a).

The angular distributions of the dimer fragmentation axis with respect to the projectile beam axis were very helpful to confirm this interpretation of the KER spectra. For peaks 2 and 3, angular distributions were found to be isotropic [6]. This is in perfect agreement with a one-site collision process for which the relative orientation of the dimer is expected to have no or very little influence.

By contrast, events associated to the peak 1 (KER between 3.5 and 5 eV) and thus to a 2-step process are strongly favored for dimer orientations quasi-parallel to the beam axis. Considering the large internuclear distance of the helium dimer compared to impact parameters leading to SC and SI, this strong dependence on the dimer orientation can be easily understood as the projectile trajectory has to pass close to both sites. A detailed analysis of the angular distribution can then provide an indirect measurement of the impact parameter dependence $P(b)$ of the SC and SI processes, assuming that the 2-step process can be considered as two independent collisions with two distinct atomic helium targets. This method gives access to information that would be inaccessible otherwise, as the transferred or emitted electron carries a significant part of the momentum transfer for such collisional energies. In Figures 9.8(a) and 9.8(b), the experimental angular distributions of peak 1 for TI and DC are compared to differential cross-sections calculated using atomic $P(b)$ probabilities for SC and SI processes. The $P(b)$ distributions were obtained using an effective single-particle approximation well established in the field of ion-atom collisions [20] for both SC and SI processes. He^+ projectiles were approximated as protons. For DC, two elementary processes were considered: SC with a He^{2+} projectile for the first step and SC with a proton projectile for the second. For TI, three processes are involved: SC with He^{2+} projectile and SI with a He^{2+} (first step) or with a proton projectile (second step). Angular distributions $F(\cos\theta)$ were then calculated by integrating over all possible impact parameters and internuclear distance R:

$$F(\cos\theta) = \int_0^\infty \int_0^\infty \int_{-\infty}^\infty P_1(b_a).P_2(b_b).P(R)\, dx\, dy\, dR \qquad (9.2)$$

where P_1 (b_a) and P_2 (b_b) are the elementary processes probabilities for steps 1 and 2, and $P(R)$ the probability for the internuclear distance R obtained from the KER distribution within the reflection principle. The excellent agreement between experimental data and theory in Figure 9.10 validated the $P(b)$ distributions used in the calculation for each elementary process. It also showed that considering this 2-step process with helium dimer targets as two independent collisions is a good approximation.

3.3 Enhanced production of low energy electrons from dimer targets

Low energy electrons[10] have been shown to efficiently cause double strand breaks in DNA [21, 22, 23, 24] and their production may have significant impact on cell survival rates. In ionizing ion atom collisions, it is well established that the energy distribution of the emitted electrons shows a continuous spectrum, resulting mainly from distant collisions with small energy transfers, which can be approximated by an exponential decay [25]. However, as it was shown by the study of He^{2+} collisions on helium dimers,

---

10 In the range of a few eV.

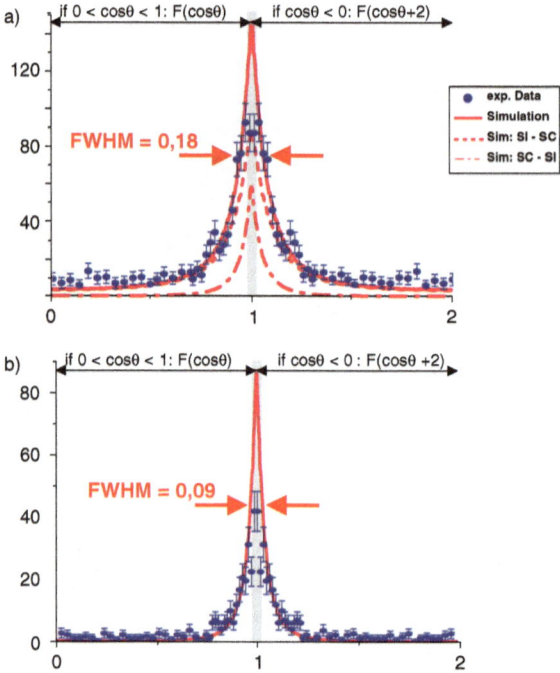

Figure 9.10: Angular distribution $F(\cos\theta)$ of the 2-step process events (3.5 eV < KER < 5 eV) for TI (a) and DC (b). The gray areas must be excluded from the comparison; they correspond to orientations with reduced detection efficiency of the experimental setup. (Reproduced with permission from [6].)

the ICD process can lead to a significant increase in low energy electron production [6]: when an excited target atom is in the vicinity of a neighbor atom, the ICD channel opens up for deexcitation of the system and results in the ionization of the latter, with emission of a low energy electron (Figure 9.9). The ICD mechanism has therefore been suggested to be a relevant process in radiation damage to biological tissue. Shortly after this first evidence for the presence of the ICD process in collisions with He_2 targets, it was shown that ICD following $He^+ + Ne_2$ collisions at 650 keV leads to an increase of electron production by a factor 14 in the $(0\,eV < E_e < 2\,eV)$[11] range [26]. This motivated a wider investigation of the ICD contribution in ion collisions with Ne_2 and Ar_2 rare gas dimer targets. New experiments were then performed with a reaction microscope, covering a large range of perturbation strength by using He ion projectiles at energies close to the maximum stopping power in water (0.125 MeV/u He^{1+}, 0.1625 MeV/u He^{1+}, and 0.150 MeV/u He^{2+}) up to very fast highly charged ions (11.37 MeV/u S^{14+}) [5]. The electron energy spectra obtained for all fragmentation channels are shown in Figure 9.11. They can be decomposed into two parts: one part at low energy, resulting

---

11 E_e is here the kinetic energy of the emitted electrons.

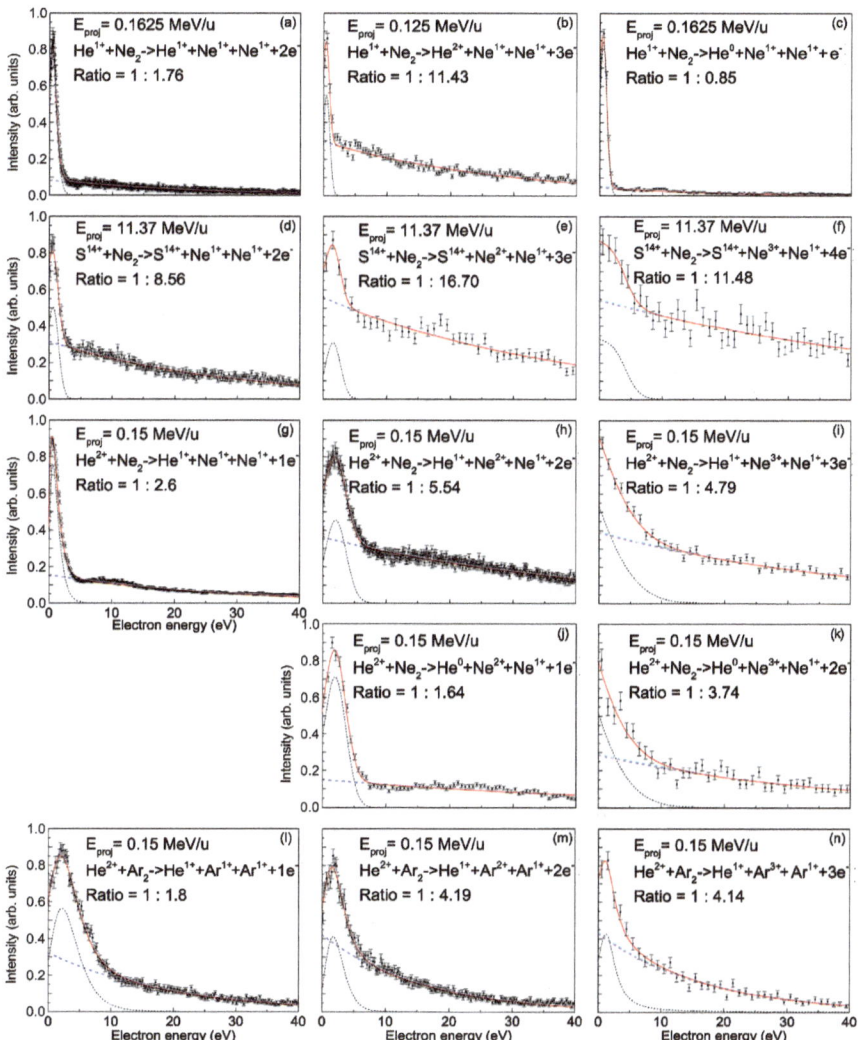

Figure 9.11: Electron energy distribution created in the observed fragmentation channels. The ratios of the ICD electrons to the continuous electrons were extracted by integrating the areas of two functions fitted to the peak (black short-dashed line) and to the exponentially decaying continuous distribution (blue long-dashed line) for the displayed electron energy range. The red solid line is the sum of the black and blue curve. (Reproduced with permission from [5].)

from ICD, and one part spreading over the whole energy range, due to direct ionization. The relative contributions of electrons resulting from ICD and of electrons emitted in the continuum by direct ionization were extracted by fitting the spectra with a Gaussian and an exponential decay function, respectively. As can be clearly seen in the Figure 9.11, ICD is responsible for an increased low-energy electron yield for a broad range of ion dimer collisions. This study suggested that in fast ion collisions with loosely bound matter, ICD is omnipresent and can be a major contributor to the emission of low-energy electrons.

ICD is usually triggered either by the creation of an inner-shell vacancy or by a multiprocess involving both ionization and excitation of the target. Such a process is thus not expected in low energy collisions ($\eta \gg 1$) with MCIs, where the dominant collision process is the capture of valence electrons. The experimental investigation of low energy collisions with Ne_2 dimer targets using 2.81 keV/u O^{3+}, 3.37 keV/u Ar^{9+} and 2.28 keV/u Xe^{20+} projectiles proved otherwise [18]. In these experiments, direct CE, RCT and ICD contributions to the fragmentation channel $(1+;1+)_F$ were extracted thanks to the KER spectra and to the charge analysis of the scattered projectile. The relative yields obtained for the three different processes are shown and compared to MC-COBM calculations in Figure 9.12.

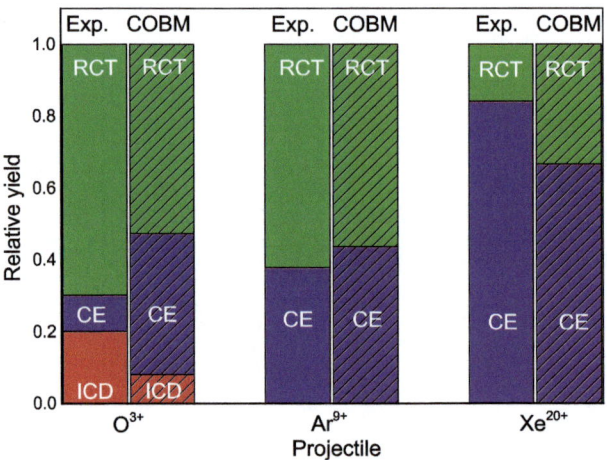

Figure 9.12: Experimental and calculated relative yields of the CE, RCT, and ICD processes contributing to the $Ne^+ + Ne^+$ fragmentation channel. Experimental uncertainties, not indicated here, were estimated to be about 0.05. (Reproduced with permission from [18].)

For the low charged O^{3+} projectiles, the ICD process represents about 20 % of the events leading to the $(1+;1+)_F$ fragmentation channel. It is far from being negligible, and the ICD contribution was indeed found to produce about 10 times more electrons than Auger decay following double capture [18]. The ICD decay channel is attributed here to a single electron capture of a $2s$ electron. This appearance of inner-shell capture for O^{3+} projectiles as well as the relative CE and RCT contributions for the two other collision systems is quite well reproduced by the model calculations. One can note however that the calculated relative yields of the three process for O^{3+} projectiles do not fairly match the data. This discrepancy shows the limits of the MC-COBM calculations for low projectile charge states, where the hydrogenic approximation becomes too rudimentary to provide precise capture probabilities. Nevertheless, with O^{3+} projectiles the model does predict a significant contribution of inner-shell capture whose main relaxation mechanism is expected to be through ICD. Calculations performed for O^{q+} projectiles ($q = 2$ to 5) have also shown that the inner shell capture contribution, and thus the ICD relative electron yield, strongly depends on the charge, with only ~1% for $q = 5$, ~3% for $q = 4$, ~8% for $q = 3$, and ~30% for $q = 2$. This dependence cannot be explained by the probability to transfer a $2s$ electron to the projectile which increases with the projectile charge state. The explanation simply arises from the fact that, for higher projectile charges, inner shell capture becomes systematically followed by the capture of additional electrons and ends up in multiple capture channels that do not contribute to the ICD process [18].

This work has shown that inner-shell capture followed by ICD can also be significant in slow collisions with low charge projectiles. It is a new and unexpected source of low energy electrons in low energy collisions that strongly overtakes Auger electron emission in the case of $O^{3+} + Ne_2$ collisions. This analysis confirmed that ICD has to be recognized as a "universal" relaxation mechanism that can play an important role in any collision involving atoms in weak interaction with a chemical environment.

4 A classical model for dimer targets

In Section 3.2, we have shown that, in fast collision between He^{2+} ions and He_2 targets, the collision could lead to a single interaction with one site of the dimer or to a 2-step process involving sequential interactions with both He atoms. In the latter case, calculations performed by considering the dimer as two independent atomic targets were found in good agreement with the differential cross-sections in angular distribution obtained experimentally. This approximation was quite natural for this collision system where the internuclear distance between the two atomic sites is large compared to the impact parameters leading to ionization or charge exchange and compared to the electronic wave function extension. For low energy collisions between highly charged ions and dimer targets such as Ar_2 and Ne_2, the situation is quite different: impact parameters leading to charge exchange (of the order of 10 a. u.) are comparable to the internuclear distance and the collision cannot anymore be considered as a 2-step process. On the other hand, a full quantum-mechanical treatment of the collision between a MCI and the dimer target is impossible in practice, and approximation methods must be used. For the charge exchange process with atomic targets, several classical approaches where successively developed [27, 28]. For low energy collisions, the classical over-the-barrier model (COBM) [29, 30, 31] was progressively achieved in the eighties. This model is based on the assumption that electrons can transit from the target to the projectile at given internuclear distances for which the height of the potential barrier between the two nuclei becomes lower than the Stark shifted binding energy of the electrons. In its most advanced version by Niehaus, the model was proven to be successful to predict charge exchange cross-sections [31, 32] as well as post-collision processes [33, 34] for atomic collisions with low energy[12] MCIs (projectile velocities v_p from 0.01 a. u. to 1 a. u.).

For molecular targets, a three-center COBM model based on the earlier work of Barany et al. [30] has been recently developed by Ichimura and Ohyama-Yamaguchi [35]. Applied to N_2 molecular targets, the model was found to be consistent with the experimental data of [19], with a preference for dissociation into quasi-equally charged fragments [36]. With rare gas dimer targets, this model could qualitatively reproduce the charge asymmetry in the fragment pair distribution when adding a specific screening parameter [37].

A quite similar approach was used soon after by Iskandar et al., [16] with the MC-COBM (Monte-Carlo classical over-the-barrier model). Within this three-center model, the dimer target is considered as two quasi-independent atomic targets fixed in space: the projectile can interact with both targets at the same time, but direct interactions between the two atomic sites are neglected. The MC-COBM also includes additional ingredients. As proposed by Niehaus for atomic targets [31], this model comprises two distinct stages, the way in, where electrons of the target become shared with the projectile, and the way out, where they end up captured by the projectile or recaptured by the target. In order to facilitate both the theoretical treatment and the comparison with the experimental data, Monte Carlo (MC) simulations replace a full analytical calculation. These developments allow the prediction of the final ion pair production, give access to capture multiplicity on each site as a function of the impact parameter **b** in the molecular frame, and provide the transverse momentum exchange between the projectile and each center of the dimer all along the interaction path. As seen in the previous sections, calculations performed using this model were found in reasonable agreement with the experimental data obtained in low energy collisions between MCIs and dimer targets [17, 18, 16]. A brief description of the model and complementary tests are given in the following.

4.1 Reminder of the COBM for atomic targets

A schematic view of the interaction between a projectile A^{q+} and an atomic target B is shown on the Figure 9.13. In the way in, as the projectile is approaching the target, the potential barrier decreases and electrons from the target become shared with the projectile at specific internuclear radii R_i^{in}. For an electron numbered i (starting from the least bound electron to the most bound electron) this sharing radius depends on

---

12 At higher velocity, ionization and excitation are not anymore negligible.

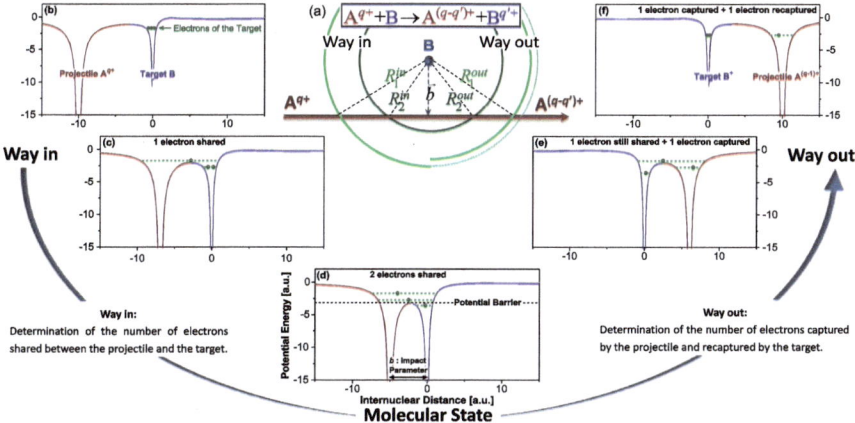

Figure 9.13: (a) Schematic view of the interaction between a projectile A^{q+} and a target B with N_B shared electrons involved on the way in ($N_B = 2$ in this example). (b–f) Illustration of the different steps leading to the sharing of two electrons in the way in and to the capture of one electron in the way out. See text for details. (Reproduced with permission from [16].)

its binding energy to the target I_i^B, and on the projectile and target effective charge $q_i^{A_{in}}$ and $q_i^{B_{in}}$:

$$R_i^{in} = \frac{i + \sqrt{q_i^{A_{in}} q_i^{B_{in}}}}{I_i^B} \tag{9.3}$$

In the COBM model as proposed by Niehaus, shared electrons are transferred to molecular states where they do not contribute to the charge screening between the projectile and the target.[13] In the way in, we thus use a constant projectile effective charge $q_i^{A_{in}} = q$ while the target effective charge $q_i^{B_{in}}$ increases by one unit each time one electron is shared. The total number N_B of electrons being shared depends on the impact parameter b, where the condition $b < R_{N_B}^{in}$ must be fulfilled.

On the way out, as the projectile goes away from the target, the Coulomb barrier height increases with the internuclear separation between the projectile and the target. When the Coulomb barrier reaches the binding energy of a shared electron i, at the distance R_i^{out}, this electron is then either captured by the projectile or recaptured by the target. R_i^{out} is given here by

$$R_i^{out} = R_i^{in} \frac{(\sqrt{q_i^{A_{out}}} + \sqrt{q_i^{B_{out}}})^2}{(\sqrt{q_i^{A_{in}}} + \sqrt{q_i^{B_{in}}})^2} \tag{9.4}$$

---

13 In this classical picture, shared electrons remain in the peripheral area as the projectile gets closer to the target nucleus.

where $q_i^{A_{out}} = q - c_i$ and $q_i^{B_{out}} = i + c_i$ are the effective charges of the projectile and target in the way out, with c_i the number of electrons previously captured by the projectile. The probabilities P_i to be captured by the projectile and $(1 - P_i)$ to be recaptured by the target are assumed to only depend on the multiplicity of quantum states available on each site. P_i is approximated by

$$P_i = \frac{n_i^2}{m_i^2 + n_i^2} \tag{9.5}$$

where n_i and m_i are effective principal quantum numbers estimated in the framework of the Bohr model. The details of their calculation can be find in [31, 16]. In the description given by Niehaus, the model finally provides cross-sections for each series of events leading to a given capture multiplicity, by integrating over the impact parameters with the associated capture probabilities [31].

4.2 A classical model for dimer targets

In the MC-COBM, the same approach as for a simple atomic target can be used, by considering the collision with the two centers of the dimer as two separate collisions with two atoms. This approximation is justified by the low electron mobility within the dimer that was previously discussed in Section 3.1. The only indirect effect of one site on the other is due to the change of the effective charge of the projectile during the collision, as this change can be caused by captured electrons from both atomic sites. This indirect influence as well as the additional degrees of freedom due to the geometry of a diatomic dimer target are the major motivations to use a MC simulation. The MC simulation allows to sample randomly all possible orientations of the dimer target in 3D space, and all possible impact parameters possibly leading to charge exchange (see [16] for details). The internuclear distance between the two sites is simply fixed to its average value. The COBM model is then applied as described in the previous section, with two targets noted B and C. For each given impact parameter vector **b** in 3D space and each orientation of the dimer provided by the MC simulation, the sequence of intersections of the projectile trajectory with sphere of radii $R_i^{B_{in}}$ and $R_i^{C_{in}}$ centered on centers B and C is computed, leading to the maximum number of electrons N_B and N_C being potentially shared in the way in. $R_i^{B_{in}}$ and $R_i^{C_{in}}$ are calculated using equation (9.3), by replacing the subscript i by the number of electrons i_C and i_B that are shared by each atomic center. Similarly, when entering the way out of each atomic center, the sequence of intersections with spheres of radii $R_i^{B_{out}}$ and $R_i^{C_{out}}$ is calculated using equation (9.4). The effective charge of each target site remains the same as for the atomic case. For the target B (C),

$$q_{i_{B(C)}}^{B(C)_{in}} = i_{B(C)} \quad \text{and} \quad q_{i_{B(C)}}^{B(C)_{out}} = i_{B(C)} + c_{i_{B(C)}} \tag{9.6}$$

By contrast, the projectile effective charge becomes:

$$q_{i_{B(C)}}^A = q - c_{i_B} - c_{i_C} \qquad (9.7)$$

both for the way in and the way out, where c_{i_B} and c_{i_C} are the number of electrons previously captured from sites B and C. Because of the different possible orientations of the dimer, molecularization, and capture of electrons from B and C can occur in different order. This implies that the evaluation of the crossing radii and the determination of capture or recapture of the electrons has to be done sequentially to know at any step i_B, i_C, c_{i_B}, and c_{i_C}, as the projectile progresses along the collision axis. At the end of the collision, we keep track of the number of electrons captured from each center of the dimer. The number i_B and i_C of the individual captured electrons are recorded, giving access to their initial atomic shell [18] as well as the initial coordinates of the projectile and of the dimer sites B and C for the study of impact parameter dependence [17].

Using MC simulations shows another advantage: as the distances between the projectile and the two target sites as well as their effective charges are known for every step, one can integrate the momentum exchange between the projectile and the dimer due to Coulomb repulsion all along the trajectory, as detailed in [16]. The transverse component of the momentum exchange is directly linked to the scattering angle of the projectile, and within the present classical approximation, it is strongly correlated to the impact parameter. It provides additional points of comparison with experimental data and further insight into the collision process.

4.3 Comparison with experimental results

Electron capture cross-sections predicted by the MC-COBM model have been compared to experiment data obtained with O^{3+}, Ar^{9+} and Xe^{20+} projectiles (at velocities ranging from 0.3 a. u. to 0.4 a. u.) colliding on both Ne_2 and Ar_2 dimer targets. For these collision systems, the relative yields of the different capture channels, corresponding to a given number of electrons captured on each site, have been found in good agreement with almost all the data, showing deviations that remained below 10 % of the total yield [18, 16]. Significant discrepancies were only observed for collisions with low charged O^{3+} projectiles, where the hydrogenic approximation of the model does not hold anymore [18].

As shown in Section 3.2, the correlations between the orientation of the dimer target and the direction of the scattered projectile that were calculated for $Ar^{9+} + Ar_2$ collision were also found in excellent agreement for all the capture channels. Thanks to the MC-COBM model, a realistic picture of the impact parameters, in the molecular frame, leading to the different capture channels could be obtained [17].

Another study was recently dedicated to the comparison of predicted differential cross-sections (DCS) in transverse momentum exchange with experimental data obtained using Ar^{9+} and Xe^{20+} projectiles colliding on Ne_2 and Ar_2 dimer targets [16]. As the projectile scattering angle is very sensitive to the effective charges of the collision partners during the collision, several scenarios using different charge screening parameters S were used. For the sake of simplicity, only the screening of the target was investigated. The effective charges of sites $B(C)$ respectively become

$$q_{i_{B(C)}}^{B(C)_{in}} = i_{B(C)} \times (1-S) \quad \text{and} \quad q_{i_{B(C)}}^{B(C)_{out}} = i_{B(C)} \times (1-S) + c_{i_{B(C)}} \tag{9.8}$$

In the scenario corresponding to the initial description by Niehaus with $S = 0$, shared electrons are considered to stay on outer orbitals and do not participate to screening. Calculations were also performed assuming a full screening of the shared electrons ($S = 1$), and for an intermediate case with $S = 1/2$. The DCS obtained for $S = 1/2$ were found to agree remarkably well with all the experimental data. An example is given in Figure 9.14 for collisions using Ar^{9+} projectiles.

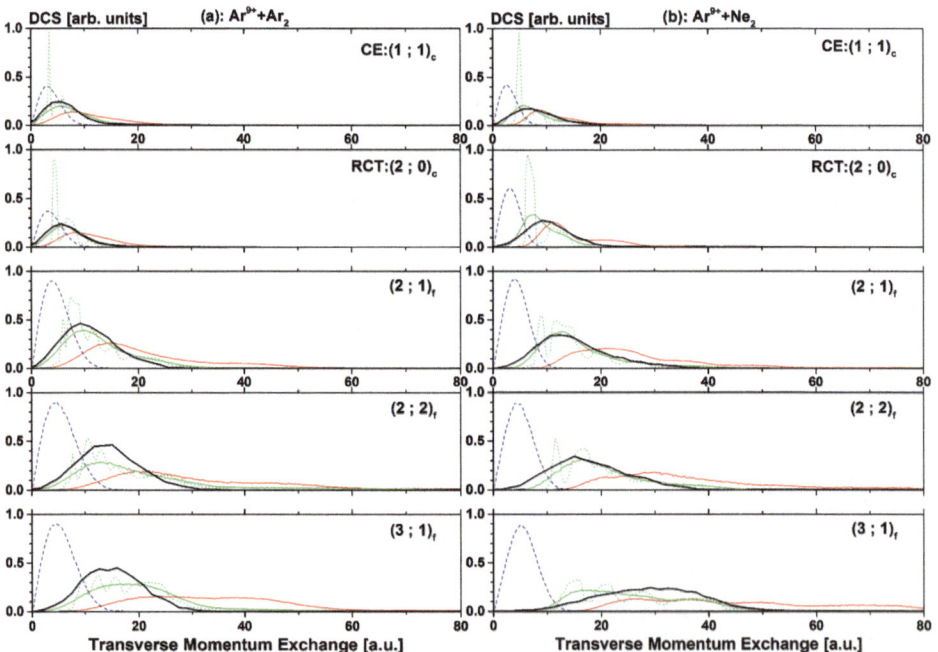

Figure 9.14: Differential cross-section (DCS) in transverse momentum exchange obtained experimentally (thick black line) and using the MC-COBM for Ar^{9+} projectiles colliding with Ar dimers (a) and Ne dimers (b). The MC-COBM distributions, previously normalized to the experimental data, are given for three different charge screening parameters: $S = 1$ (blue dashed line), $S = 1/2$ (green line) and $S = 0$ (red line). For $S = 1/2$, the distribution prior convolution with the apparatus response function is also given as a dotted green line. (Reproduced with permission from [16].)

Despite its simplicity, the MC-COBM has proven to be very efficient for the prediction of relative capture cross-sections. Even for differential cross-sections, whose calculations are usually challenging, the model was found in excellent agreement with the experiment when using target charge screening. The latter point indicates that shared electrons are not purely spectators of the collision as it is assumed in the Niehaus model.

5 Perspectives

With a few experimental campaigns performed within only a few years, our under-standing of ion-dimer collisions has increased considerably. The low electron mobility between the two sites of the target, characteristic of van der Waals bound molecules, has been shown to be responsible for behaviors contrasting strongly with the usual relaxation of covalently bound molecular ions. In particular, it involves very little re-arrangement of the charge repartition within the dimer after the collision, which gives almost direct access to the primary collision process. As shown in [6, 16], dimer targets are thus good candidates to benchmark collision models developed for structured tar-gets and study the effect of a neighboring atom on the primary collision process. Such weakly bound systems were also found to be an ideal laboratory to study the effect of the environment on how the deposited energy due to ion impact dissipates. Both at low and intermediate collision energy, it was shown that quite unexpected relaxation processes such as radiative charge transfer and interatomic Coulombic decay play a major role. The enhanced production of low-energy electrons due to the ICD process, usually neglected in the modeling of ion-matter interactions at a microscopic level, could have a significant influence on cell survival rates after exposure to ion-radiation.

The RIMS technique is found to be particularly well adapted to the experimental investigation of collisions with dimers. Allowing the detection of several ionic frag-ments from a target, its limitation in fragment multiplicity is only due to the cross-sections of the processes of interest and to the absolute efficiency of the detectors. The recent developments of high efficiency micro-channel plates[14] will open the path to the study of collisions with more complex targets, leading to the emission of 3, 4 or more charged fragments. Photoionization experiments involving dimers of small covalent molecules [38, 39] and small rare gas clusters [40] have already been per-formed, but the investigation of ion collisions with such targets is just emerging. In a pioneering experiment with $(N_2)_2$ targets, the three-body fragmentation channels of $(N_2)_2^{(3,4)+}$ molecular ions were used to evaluate the role of a neighbor in the fragmenta-tion dynamics of dimers made of two covalent molecules [41]. These studies constitute additional steps on the bridge between the few body problem and ion-interaction with complex matter. Small clusters, and more particularly covalent molecule dimers, are

---

14 Microchannel plates are the main elements of the position sensitive detectors used in RIMS exper-iments. Their absolute efficiency for ion detection has recently been increased up to 90 %.

also a subject of interest *per se*: diatomic molecular dimers can have different geometries such as the so-called T, X, H, and S structures (see [42] and references therein) depending on the relative position and orientation of the two molecules constituting the dimer. The Coulombic explosion of a complex target following its multiionization by a MCI projectile, the so-called Coulomb imaging, can serve as a direct probe providing measurements of interatomic or intermolecular distances and of the relative orientation of the molecules. Such information will surely complete our knowledge in a close future and provide stringent tests of ab-initio calculations of potential energy surfaces in quantum chemistry [42].

The study of larger molecular clusters is another challenge that may also someday become at reach. New processes, that are too slow to be observed in the fast breakup of a free molecule, could then be revealed in the relaxation of molecules embedded in a matrix. If the execution of such experiments remains extremely difficult, its outcome would bring us even closer to a complete understanding of the interactions between ions and complex matter.

Bibliography

[1]   Grisenti RE et al. Determination of the Bond Length and Binding Energy of the Helium Dimer by Diffraction from a Transmission Grating. Phys. Rev. Lett. 2000;85:2284.
[2]   Ullrich J, Moshammer R, Dorn A, Dörner R, Schmidt L, Schmidt-Böcking H. Recoil-ion and electron momentum spectroscopy: reaction-microscopes. Rep. Prog. Phys. 2003;66:1463.
[3]   Dörner R, Mergel V, Jagutzski O, Spielberger L, Ullrich J, Moshammer R, Schmidt-Böcking H. Cold Target Recoil Ion Momentum Spectroscopy: a 'momentum microscope' to view atomic collision dynamics. Phys. Rep. 2000;330:95.
[4]   Matsumoto J et al. Asymmetry in Multiple-Electron Capture Revealed by Radiative Charge Transfer in Ar Dimers. Phys. Rev. Lett. 2010;105:263202.
[5]   Kim H-K et al. Ion-impact-induced interatomic Coulombic decay in neon and argon dimers. Phys. Rev. A. 2013;88:042707.
[6]   Titze J et al. Ionization dynamics of helium dimers in fast collisions with He++. Phys. Rev. Lett. 2011;106:033201.
[7]   Matsumoto J et al. Multiple-ionization and dissociation dynamics of a rare gas dimer induced by highly charged ion impact. Phys. Scr. T. 2011;144:014016.
[8]   Jahnke T et al. Experimental Observation of Interatomic Coulombic Decay in Neon Dimers. Phys. Rev. Lett. 2004;93:163401.
[9]   Morishita Y et al. Experimental Evidence of Interatomic Coulombic Decay from the Auger Final States in Argon Dimers. Phys. Rev. Lett. 2006;96:243402.
[10] Cederbaum LS, Zobeley J, Tarantelli F. Giant Intermolecular Decay and Fragmentation of Clusters. Phys. Rev. Lett. 1997;79:4778.
[11] Walsh B et al. Electron capture from C60 by slow multiply charged ions. Phys. Rev. Lett. 1994;72:1439.
[12] Ben-Itzhak I et al. Multiple-electron removal and molecular fragmentation of CO by fast F4+ impact. Phys. Rev. A. 1993;47:2827.

[13] Wohrer K et al. Dissociation of multicharged CO molecular ions produced in collisions with 97-MeV Ar14+: Dissociation fractions and branching ratios. Phys. Rev. A. 1992;46:3929.
[14] Remscheid A, Huber BA, Pykavyj M, Staemmler V, Wiesemann K. Electron capture and dissociation of the N2q+ molecule in slow Ar8+/N2 collisions. J. Phys. B. 1996;29:515.
[15] Cassimi A et al. In: Ullrich J, Shevelko VP, editors. Many-Particle Dynamics in Atomic and Molecular Fragmentation. Berlin, Heidelberg, New York: Springer Verlag; 2003.
[16] Iskandar W et al. Coulomb over-the-barrier Monte Carlo simulation to probe ion-dimer collision dynamics. Phys. Rev. A. 2018;98:012701.
[17] Iskandar W et al. Atomic Site-Sensitive Processes in Low Energy Ion-Dimer Collisions. Phys. Rev. Lett. 2014;113:143201.
[18] Iskandar W et al. Interatomic Coulombic Decay as a New Source of Low Energy Electrons in Slow Ion-Dimer Collisions. Phys. Rev. Lett. 2015;114:033201.
[19] Ehrich M, Werner U, Lutz HO, Kaneyasu T, Ishii K, Okuno K, Saalmann U. Simultaneous charge polarization and fragmentation of N2 molecules in slow keV collisions with Kr8+ Ions. Phys. Rev. A. 2002;65:030702(R).
[20] Kirchner T, Lüdde HJ, Horbatsch M. A time-dependent quantal approach to electronic transitions in atomic collisions. Recent Res. Devel. Physics. 2004;5:433.
[21] Boudaïffa B, Cloutier P, Hunting D, Huels MA, Sanche L. Resonant formation of DNA strand breaks by low-energy (3 to 20 eV) electrons. Science. 2000;287:1658.
[22] Hanel G, Gstir B, Denifl S, Scheier P, Probst M, Farizon B, Farizon M, Illenberger E, Märk TD. Electron Attachment to Uracil: Effective Destruction at Subexcitation Energies. Phys. Rev. Lett. 2003;90:188104.
[23] Sanche L. Low energy electron-driven damage in biomolecules. Eur. Phys. J. D. 2005;35:367.
[24] Martin F, Burrow PD, Cai Z, Cloutier P, Hunting D, Sanche L. DNA Strand Breaks Induced by 0–4 eV Electrons: The Role of Shape Resonances. Phys. Rev. Lett. 2004;93:068101.
[25] Rudd ME, Toburen LH, Stolterfoht N. Differential cross sections for ejection of electrons from argon by protons. At. Data Nucl. Data Tables. 1979;23:405.
[26] Kim H-K et al. Enhanced production of low energy electrons by alpha particle impact. Proc. Natl. Acad. Sci. USA. 2011;108:11821.
[27] Bohr N, Lindhard J, Dan K. Vidensk. Selsk. Mat. Fys. Medd. 1954;28:7.
[28] Knudsen H, Haugen HK, Hvelplund P. Single-electron-capture cross section for medium- and high-velocity, highly charged ions colliding with atoms. Phys. Rev. A. 1981;23:597.
[29] Ryufuku H, Sasaki K, Watanabe T. Oscillatory behavior of charge transfer cross sections as a function of the charge of projectiles in low-energy collisions. Phys. Rev. A. 1980;21:745.
[30] Barany A et al. Absolute cross sections for multi-electron processes in low energy $Ar^{q+} - -Ar$ collisions: Comparison with theory. Nucl. Instr. Meth. B. 1985;9:397.
[31] Niehaus A. A classical model for multiple-electron capture in slow collisions of highly charged ions with atoms. J. Phys. B. 1986;19:2925.
[32] Niehaus A. Spontaneous electron emission from slow atomic collisions. Phys. Rep.. 1990;186:149.
[33] Niehaus A. Multiple electron capture in slow collisions of highly charged ions with atoms. Nucl. Instrum. Meth. B. 1988;31:359.
[34] Guillemot L, Roncin P, Gaboriaud MN, Laurent H, Barat M. Critical study of the molecular Coulombic barrier model for multiple electron capture by highly charged ions. J. Phys. B. 1990;23:4293.
[35] Ohyama-Yamaguchi T, Ichimura A. Near-site far-site charge asymmetry in diatomic Coulomb fragmentation with slow highly charged ions. Nucl. Instr.Methods B. 2005;235:382.
[36] Ohyama-Yamaguchi T, Ichimura A. Analysis of charge-asymmetric Coulomb explosion of N2 molecules with slow Kr8+ ions. J. Phys. Conf. Ser. 2009;163:012047.

[37] Ohyama-Yamaguchi T, Ichimura A. Multiple ionization of rare gas dimers by slow highly charged ions: screening effect during a collision. Phys. Scr. T. 2013;156:014043.

[38] Jahnke T et al. Ultrafast energy transfer between water molecules. Nature Physics. 2010. doi:10.1038/NPHYS1498.

[39] Ding Xiaoyan, Haertelt M, Schlauderer S, Schuurman MS, Naumov AYu, Villeneuve DM, McKellar ARW, Corkum PB, Staudte A. Ultrafast Dissociation of Metastable CO2+ in a Dimer. Phys. Rev. Lett. 2017;118:153001.

[40] Ulrich B et al. Imaging of the Structure of the Argon and Neon Dimer, Trimer, and Tetramer. J. Phys. Chem. A. 2011;115:6936.

[41] Méry A et al. Role of a Neighbor Ion in the Fragmentation Dynamics of Covalent Molecules. Phys. Rev. Lett. 2017;118:233402.

[42] Gomez L, Bussery-Honvault B, Cauchy T, Bartolomei M, Cappelletti D, Pirani F. Global fits of new intermolecular ground state potential energy surfaces for N2–H2 and N2–N2 van der Waals dimers. Chem. Phys. Lett. 2007;445:99.

Index